Broadcast Data Systems

Teletext and RDS

Related titles from Butterworths

Broadcast Sound Technology
Michael Talbot-Smith

Lighting Control Systems for Television, Film and Theatre
Jack Kelleher

Loudspeaker and Headphone Handbook
Edited by John Borwick

The RTS Television Engineering Handbook

TV and Video Engineer's Reference Book
Edited by G. B. Townsend and K. G. Jackson

. . . and from Focal Press

The Art of Digital Audio
John Watkinson

The Art of Digital Video
John Watkinson

Coding for Digital Recording
John Watkinson

Microphones: Technology and Technique
John Borwick

Stereo Sound for Television
Francis Rumsey

Broadcast Data Systems
Teletext and RDS

Peter L. Mothersole FEng, CEng, FIEE

Norman W. White CEng, FIEE

Butterworths
London Boston Singapore Sydney Toronto Wellington

PART OF REED INTERNATIONAL P.L.C.

First published 1990

British Library Cataloguing in Publication Data
Mothersole, Peter L. (Peter Leonard)
Broadcast data systems: teletext and RDS.
1. Teletext systems
I. Title II. White, Norman W.
621.388
ISBN 0-408-04815-8

Library of Congress Cataloguing in Publication Data
Mothersole, Peter L.
Broadcast data systems: Teletext and RDS/
Peter L. Mothersole. Norman W. White.
p. cm.
Includes bibliographical references.
ISBN 0-408-04815-8
1. Teletext systems. I. White, Norman W.
II. Title.
TK5105.M84 1990
384.3'52—dc20 90-33586

Photoset by Genesis Typesetting, Laser Quay, Rochester, Kent
Printed and bound by Hartnolls Ltd, Bodmin, Cornwall

Foreword

It is now some 20 years since the progress of semiconductor technology, stimulated by the computer industry, first made it realistic to contemplate the addition of data processing devices to a product as sensitive to cost as the television receiver. An enormous potential market was seen, built on the existing mass TV set market. Another spur for rapid development was the spirit of competition between the UK broadcasters – the BBC and the IBA. This period quickly led the broadcasters to give separate 'over-air' public demonstrations of their technically different systems in 1973. Then, working with industry and government in a joint technical committee under BREMA's chairmanship, a unified specification, soon dubbed 'The Teletext Spec', was agreed and published in 1974. Separate BREMA/Post Office talks led to the adoption of the same display standard for 'Viewdata' (later 'Prestel').

It had been evident to the members of the joint committee – of whom one was Peter Mothersole – that teletext would have wider applications than the broadcasting of textual information and subtitles. Provision was therefore made so that later extensions could be introduced without making earlier teletext equipments obsolete. This 'backward compatibility' has proved invaluable in allowing new applications to be grafted onto the specification as needed. An example is the 1990 extension to provide for the automatic control of videocassette recorders.

The extended UK teletext system is now known internationally as CCIR Teletext System B or World System Teletext (WST). It has now been adopted by most of the countries which operate a teletext service. This book is therefore of immediate value in many countries throughout the world. It describes in some detail, in a readable, non-mathematical style, the systems and techniques needed to provide a comprehensive data broadcasting service. The direct approach to the applications described arises from the authors' first-hand experience over many years. The book should therefore be of use to newcomers to the topic who are anxious to get on-air as quickly and efficiently as possible; to those wishing to widen their experience; and to students whose privilege it may be in the future to build on what is described here, to achieve yet better things.

George McKenzie
Formerly Head of the Automation and Control Section,
IBA, Crawley Court

Preface

Teletext and the Radio Data System provide digital communication channels that are combined with the normal television and VHF radio signals. The origination and transmission of these digital signals introduce new equipment and techniques for their implementation in broadcasting systems. This book is intended to bridge the gap between established television and radio practice and the computer technology that is used in these new systems for practising engineers and students.

We gratefully acknowledge the use of material published by the BBC, the IBA, the IRT and the EBU and for the assistance given by Mr B. J. Rogers and our colleagues at VG Electronics Limited.

P.L.M.
N.W.W.

Contents

1 The development of teletext

Teletext is a system of transmitting digitally coded alphanumeric data in the field blanking interval (FBI) of a television signal without disturbing the normal vision or sound signals. The data signal is decoded in the receiver and displayed as a page of information on the screen as an alternative to the video picture. Various methods for utilizing the spare time in the field blanking interval have been proposed as a means of transmitting additional information to the home, but it was not until about 1970, when Random Access Memory (RAM) and Read Only Memory (ROM) integrated circuits (ICs) became available, that practical systems become potentially viable.

The BBC Research Department had at that time been studying methods of subtitling programmes for the hearing impaired, which, in isolation, did not seem to be an economic proposition. However, the availability of larger memories made it feasible to expand the system so that many pages of information could be transmitted which would then extend the scope of the system and make it useful to all viewers. This proposed system was called CEEFAX. The IBA had for some time been using digital data carried in the FBI for the transmission of information in the television network, but this data was not normally broadcast. As a result of development work undertaken by the IBA the system was extended to include pages of information that could be broadcast in an arrangement similar in principle to that proposed by the BBC; and was called ORACLE.

In 1972 both of the UK Broadcasting Authorities had proposed these systems for transmitting information in digitally coded form in the FBI, but the systems had rather different technical parameters. Development work was undertaken which involved both of the Broadcasting Authorities and a number of receiver manufacturers, notably Philips, GEC and Rank Radio International. This work was coordinated by a BREMA (British Radio Equipment Manufacturers' Association) Working Group. The Group produced the specification for a 'unified system', the parameters of which were first published in 1974[1]. Extensive field tests had been undertaken during the development of the unified system and these were continued in order to gain experience and also to increase confidence in the proposals. Further work was undertaken by receiver designers on various aspects of reception and decoding of data signals with high data rates. The influence of the frequency and phase response of IF amplifiers and video detector circuits in teletext receivers was studied and the performance of existing receivers evaluated.

A viable teletext service would need to have a service area substantially similar to that of the television service. The broadcasting authorities implemented studies which involved measurements, using receivers at fixed and mobile locations, to determine that this aim was achievable with the proposed unified system. They also established the necessary transmission standards.

The amplitude and the position of the teletext signals in the FBI had to be chosen so that no disturbance would be caused to the picture or sound reproduction on existing television receivers. The amplitude of the teletext signal was chosen so that no 'sound buzz' occurred, even on receivers with barely adequate inter-carrier sound selectivity. Sound buzz is caused by the high frequency components of the data signal interfering with the inter-carrier sound signal. If the teletext signals are inserted onto the FBI near to the field synchronizing pulses, they can show up as 'busy dots' on the vertical flyback across the picture on receivers with inadequate blanking. If they are inserted too close to the video signal they can then show up as 'busy dots' at the top of the picture when the display is slightly underscanned.

Demonstrations and tests were also carried out with non-technical viewers to gain commercial confidence in the proposed service and also to confirm the acceptability of the chosen page format. A page format consisting of 40 characters per row and 24 rows per page had been chosen, as a result of experiments, to be a satisfactory compromise between character size and the amount of information contained in a displayed page, when viewed on a normal domestic receiver. (Computer displays normally use 80 characters per row, but are viewed by an operator sitting relatively close to the screen.)

Field tests were also undertaken in Germany by the BBC and IBA in conjunction with the IRT to study the performance at VHF, with a 5 MHz bandwidth television system[2]. In 1976 a revised specification, which included an improved graphics facility, was published[3]. This specification confirmed the broadcast standards that had been verified by the extensive range of field trials.

The UK teletext specification, first published in 1974, had been evolved bearing in mind the performance of future television receiver designs, which would have improved IF response and also improved tuning arrangements. An essential feature of the proposed system was that the receiver decoding could eventually be achieved using only two or three large scale integrated (LSI) circuits so that the additional receiver cost would only be some 5% to 10%. This was essential to ensure that teletext would eventually be incorporated in all television receivers, not only the expensive ones, and hence gain widespread public acceptance. To help meet this objective, it was essential for the teletext signal to be both functional and robust so that satisfactory reception could be achieved in difficult reception areas. In addition, the system had to be capable of future expansion to accommodate languages other than English and more advanced displays incorporating high resolution graphics [4k].

The main technical features of the proposed system were:

1. Data pulses were transmitted on otherwise unused television lines during the field blanking interval (FBI) using a bit rate of 6.9375 Mbit/s (444 × nominal line frequency).
2. Each television data line carried all the information for a complete 40-character display row.
3. Each page consisted of 24 rows of 40 characters, using both upper and lower-case characters, including a special top row called the ‘page header’ carrying information for control and display purposes. The format of the transmitted data was mapped directly to the displayed page (‘fixed-format data’).
4. The system allowed a maximum of 800 decimally numbered pages to be used which could be extended using continuation pages. The pages were divided into eight magazines, each of 100 numbered pages.
5. Using one data line per field, the system allowed two full pages per second to be transmitted. Additional data lines could be used to increase the transmission rate.
6. All the data words were 8 bits in length; parity protection was used for the character data words. Hamming codes [5] were used to protect addressing and control words, permitting the correction, at the receiver, of single errors in those data words.
7. Provision was made for news flashes and subtitles.
8. Every page header carried clock-time information, for a time display.

9. Control characters were used to provide colouring and flashing of selected words.
10. A colour graphics facility was provided.

In formulating the system, a number of factors, some of them conflicting, were considered as follows:

(a) access time and the number of pages
(b) page format, content and legibility
(c) data rate and the television system bandwidth
(d) flexibility of the system, and future developments
(e) receiver options and additional cost

It was considered that a reasonably short access time should be a prime objective, but, at the same time, adequate capacity should be available to justify the cost of the system. In 1976, two lines were used per field blanking interval; hence approximately four pages were transmitted per second. The consequent transmission time for a 100-page magazine, 24 seconds, meant that the mean time for a page to be displayed, after selection, was 12 seconds. This 'access time' could be maintained when more than one magazine was used, only by employing additional data lines in the fbi.

For a page of 960 characters (24 rows of 40 characters per row), the required page store was approximately 7000 bits: well within the capacity of one LSI circuit. To simplify the receiver decoder, all the data preamble and the information for one row of characters, 360 bits in all, was transmitted in approximately 52 microseconds within one line period. Hence, the required bit rate was some 6.9 Mbit/s. That bit rate was within the capacity of an existing 5 MHz video channel, but extra care in the design of the receiver IF amplifier and demodulator was required to prevent the performance being degraded by poor pulse shape. Aerial matching was also important since short-duration reflections also degraded the data signal, even in a good reception area.

As a result of experience gained during the two-year experimental period (1974–1976), a number of minor modifications were made to improve the information presentation. These included the display of double-height characters, the addition of coloured background to discrete areas of text and a 'graphics-hold' facility to prevent discontinuities in the graphics when changing colour. Provision was also made for the system to operate with the magazines transmitted in series (one after the other), using common FBI lines, or in parallel, using separate lines. A revised specification was issued in 1976 [3].

Alternative teletext system proposals were made at about this time both in France and Canada. The French proposal, ANTIOPE [6], had a transmission format to make it compatible with data transmission over telephone lines, but the system did not take any advantage of many of the characteristics of the television signal [7]. The Canadian proposal, TELIDON [8], was designed to produce very high quality graphics; but a much more complex decoder was required and such a decoder would not be appropriate for a domestic receiver for many years. Both these systems are 'variable-format data' which required considerably more data to format the page than a 'fixed format' system. The former also requires a considerably more elaborate error correction arrangement to prevent disruption of the page format.

A fixed-data format system has a considerable economic advantage and it also provides a much higher measure of protection against potential reception errors [4f]. As a result of continued development in several countries, the UK teletext system had, by 1984, evolved into 'World System Teletext' (WST) [9,10,11]. The system is now in use in more than 30 countries throughout the world and the decoder LSI circuits are incorporated into television receivers with very little increase in cost. Whilst the teletext standard is independent of the colour television standard being used – PAL, SECAM or NTSC – it is dependent on the line frequency and video bandwidth of the television channel. A different standard was therefore developed for 525-line receivers [4j, 12]. Teletext first became established in Europe and then extended quite rapidly to many other 625-line countries. Its penetration into 525-line countries is at the moment relatively small, but this situation is likely to change now that 525-line decoder LSI circuits are becoming available.

Fixed data format, as used by WST, permits a relatively simple decoder to be used but this has not inhibited further development of the system. New enhancements include high resolution graphics, the addition of special characters required for a wide range of different languages and the transmission of still pictures.

The various enhancements have been specified as a number of different 'levels' [13]. The main features are briefly summarized as follows:

Level 1 Ninety-six character font, double height, flashing, mosaic graphics and eight colours.

Level 2 Additional data packets (non-displayed rows) and 'pseudo pages' for transmission of additional characters and non-spacing attributes; a wider choice of background

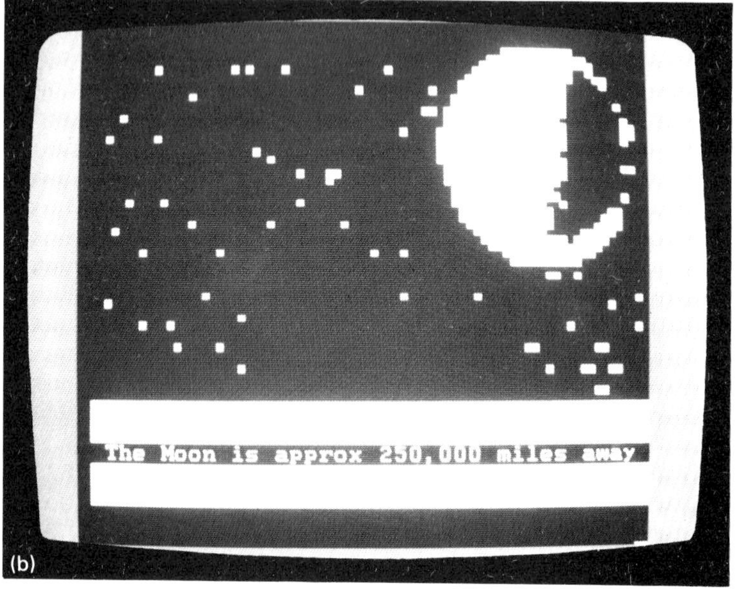

Figure 1.1 (a) Level-3 page; (b) Corresponding level-1 page

and foreground colours; a 'smoothed graphics' capability and other display features.

Level 3 DRCS (dynamically redefinable (downloaded) character sets). The character sets may include picture elements as well as special writing characters.

Level 4 Alphageometric displays. The drawing instructions assume data processing in the receiver (not needed at lower levels).

Level 5 Alphaphotographic displays – still pictures.

The higher levels, in particular levels 3, 4 and 5, require increasing amounts of page storage, and for levels 4 and 5 associated data processing, so that decoders are progressively more expensive. An example of a level-3 page is shown in Figure 1.1(a) and a corresponding level-1 page in Figure 1.1(b). Page creation for these levels requires more complex digitizers and editing software. The additional data required for each page also increases the access time, which effectively reduces the response time of the system.

The maximum number of FBI lines that can be used at present is 16 but when no video signal is present teletext data can occupy all transmitted lines, providing an increased transmission capacity of some 18 times. Such systems are called Full-field Teletext Systems.

2 The teletext data signal

Bandwidth and data rate

The teletext data and video signals are hybrid in that both analogue and digital components are present. A 1 V p–p video signal has a black level of 0 V, a sync level of −0.3 V and a brightness component which can vary in level up to a peak white value of 0.7 V. The video signal therefore consists of a generated source of mixed syncs and colour burst, plus the colour video signal.

Digital data signals which originate in computer logic circuits have fast rise times and wide bandwidths and, (normally), binary levels of +5 V and 0 V. Before the two types of signal, logic and video, can be combined, the logic data must be organized into line-length packets, and adjusted in both level and timing. The data bit rate must be as high as possible to maximize the number of data bytes per line without exceeding the video channel bandwidth. The fast rise-time logic pulses will not pass undistorted through the 5 MHz bandwidth video channel and they must be pre-shaped so that their frequency spectrum is contained within that bandwidth. Data must then be inserted cleanly onto the video signal, using a fast electronic switch controlled by an accurately timed insertion pulse.

As previously mentioned, teletext information is organized in a page format consisting of 40 characters per row and 24 rows per page. The pages are transmitted cyclicly, that is one after the other in a continuous sequence. When the complete sequence or magazine of pages has been transmitted it is then repeated. The first row of a page is referred to as the 'header row', or row '0'. This row contains the magazine and page number, the service name, date and time. In order to capture a page for display on a receiver, the magazine and page numbers are keyed in by the viewer. When the requested page in the magazine is transmitted it is then captured by the decoder and displayed on the screen.

As the pages are transmitted sequentially it is unlikely that the page will be transmitted at exactly the time a viewer requests it. There is therefore a delay between when a viewer requests a page and when the page is captured and displayed. The average delay time is called the 'access time' and is a critical parameter in a teletext system. It must be as short as possible and is related to the data rate, the data format, the number of pages being transmitted and the number of FBI lines being used for teletext.

Non-Return-to-Zero (NRZ) coding was chosen as this provides the highest bit rate for the binary coded data signal. Theoretically the minimum bandwidth required is half the bit rate. A bit rate (clock frequency) of 6.9375 Mbits per second was chosen for the 625-line systems, which have video bandwidths of 5, 5.5 or 6 MHz. The data signal frequency spectrum is centred at a frequency of 3.45875 MHz which is well inside the 625-line minimum video bandwidth of 5 MHz. The clock frequency corresponds to 444 times the horizontal line frequency of 15 625 Hz, but the data signal is not normally locked to the television signal. The bandwidth of 525-line television systems is normally 4.2 MHz and for this system a clock frequency of 5.727272 MHz was chosen, corresponding to 364 times the horizontal line frequency of 15 734.264 Hz, and again the data signal is not normally locked to the television signal.

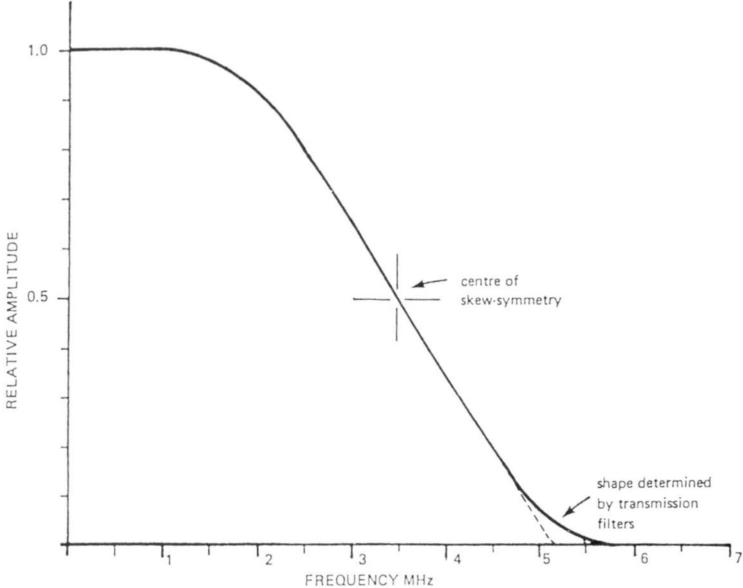

Figure 2.1 An approximate spectrum of a data pulse

The data pulse shape is chosen so that most of the energy is contained within a bandwidth of 5 MHz for 625-line systems (Figure 2.1) and 4.2 MHz for the 525-line systems. As a result of experience gained on video distribution networks feeding transmitters, a 100% raised cosine filter was found to be optimum for pulse shaping and is now standard. A typical data pulse at the input to a transmitter is shown in Figure 2.2.

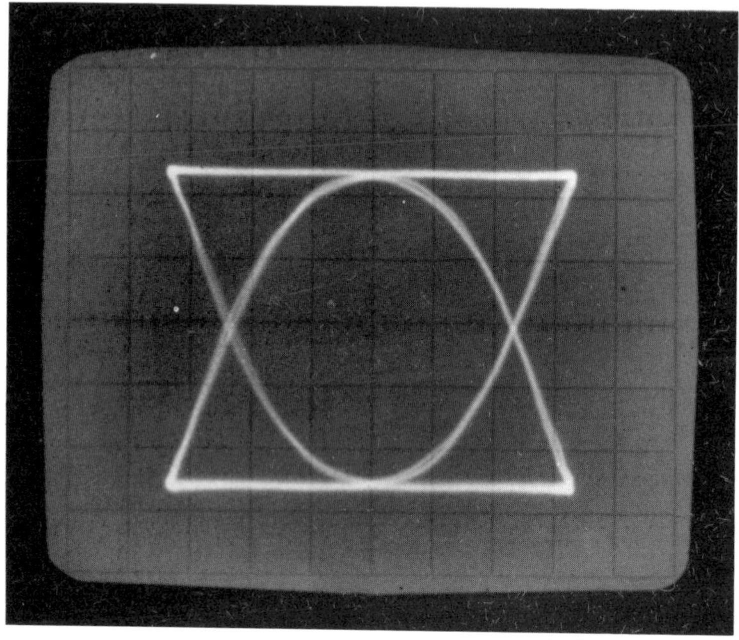

Figure 2.2 Data pulse (eye display) at a transmitter input

The data pulse amplitude is specified [13] as 66% of peak white video for WST, as shown in Figure 2.3. This amplitude was determined after extensive tests involving different receivers to ensure that the data signals did not interfere with the normal reproduction of the picture or sound.

Teletext is a 'one way' transmission system utilizing television broadcast networks. Teletext information can therefore be rapidly transmitted to millions of receivers simultaneously. To maximize the information transmission rate it is therefore desirable to minimize the amount of data required for each page. Different data formats have been specified for different teletext

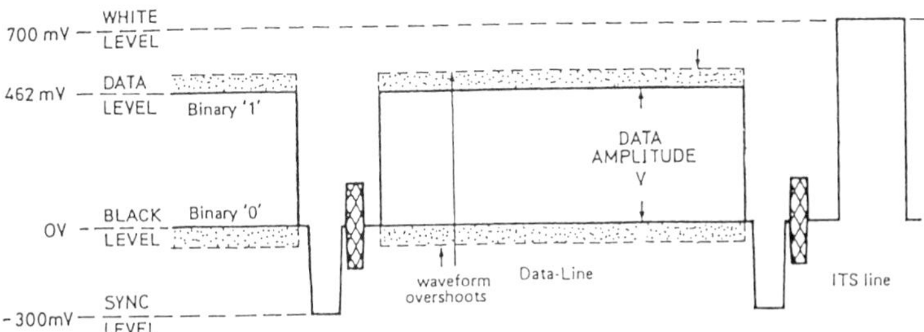

Figure 2.3 Data levels

systems [6,8,13,14], but any of these specified teletext systems can achieve similar results in such matters as the volume of displayed information and the quality of the graphics. However, there are significant differences between the systems in terms of transmission time because of the different amounts of data required to be transmitted for a given page. There are also differences in the complexity of the receiver decoders.

Teletext data format

There are basically two techniques for formatting the data for the pages that are to be transmitted in the FBI of the television signal: free-format and fixed-format.

Free-format

A page of data and graphics can be represented as a continuous string of digits using the normal 'line feed' and 'carriage return' codes to signify the end of each line. (This format is used when data is sent over telephone or data transmission lines.) A typical teletext page might contain some 8 K bits of data which could then be divided up into a series of 50-microsecond data blocks. These data blocks could be added to the fbi lines of a television signal at 50 Hz intervals, i.e. several blocks of data per field. Each block of data would need to be preceded by a data clock run-in together with some additional bits for byte synchronization. At the decoder the blocks of data would be stored so that after several field periods a complete page of information would be held in the decoder memory and could be processed for display.

With free-format data there is no relationship between the position of characters on the screen and their positions in the data block. Positional information must therefore be transmitted as part of the data stream. Any interference or distortion resulting in any loss of information during the transmission will result in errors in the received data. The errors would therefore show up as either incorrect characters in the page or information appearing in the wrong position. In order to guard against these problems various forms of error protection must be employed and this very significantly increases the amount of data required for a page of information. Furthermore a data processor is an essential part of the decoder to enable the complex data stream to be satisfactorily decoded and displayed.

Free-format is used in the French ANTIOPE system which is used in France and in the North American Broadcast Teletext Specification (NABTS) [14], which was derived from ANTIOPE and TELIDON, but has yet to be implemented as a domestic service. The advantage of a free-format system is that the data can be readily supplied from external computers as there is no fixed relationship to the television system. In theory only the bit rate need be changed as no reformatting is necessary. The disadvantage of the system is that a very significant amount of additional data is required for adequate data protection to avoid poor reception seriously disturbing the display. More complex processing is also needed to provide positional and control instructions in the decoder.

Fixed-format

The fixed-format system, as used by World System Teletext (WST) on both 625-line and 525-line systems, exploits the regular and defined timings of the television signal that carries the data to ensure that the characters are always displayed in the correct position on the screen. Very little positional information has to be transmitted and the decoder is correspondingly simplified.

One row of teletext data (40 characters) is transmitted in one television line period in the 625-line version. Each row starts with a 'run-in' data sequence to synchronize the decoder clock, followed by byte synchronizing bits for logic synchronization, and then the magazine and row number, as shown in Figure 2.4. This data is then followed by 8 bit codes corresponding to each of 40 characters in one row of displayed information. When no character is present a 'space' character is transmitted so that one row of information always has code corresponding to 40 characters present in the data stream. There is therefore a direct 'one-to-one'

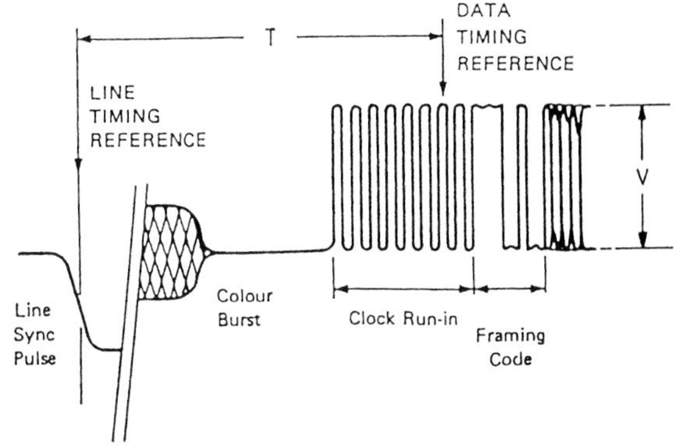

Figure 2.4 Data timing and run-in sequence

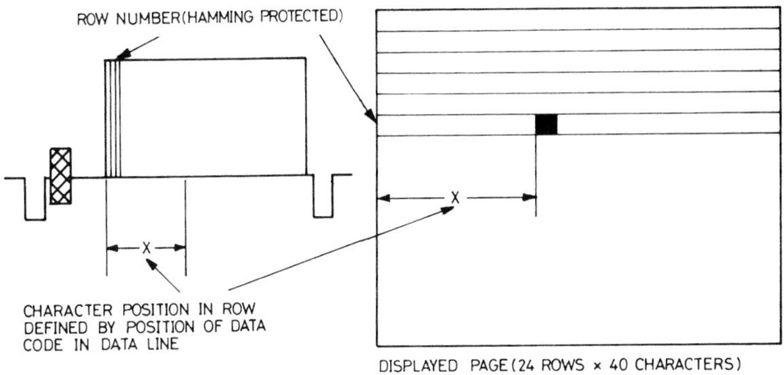

Figure 2.5 Fixed page format used by WST

correspondence between the position of the character in the display and its location in the line of data, as illustrated in Figure 2.5.

A page transmission normally starts with the top row, row 0, and finishes with the succeeding row 0 of the next page. Row 0 is called the 'header row' and contains the magazine and page number, service name, date and time. After the header row, row 1, row 2, row 3 etc., are transmitted. The page-header and row formats are shown in Figure 2.6(a) and (b).

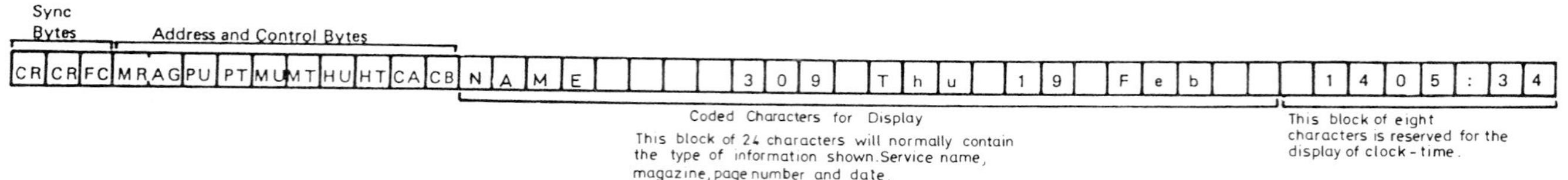

Figure 2.6(a) Page-header format (row 0)

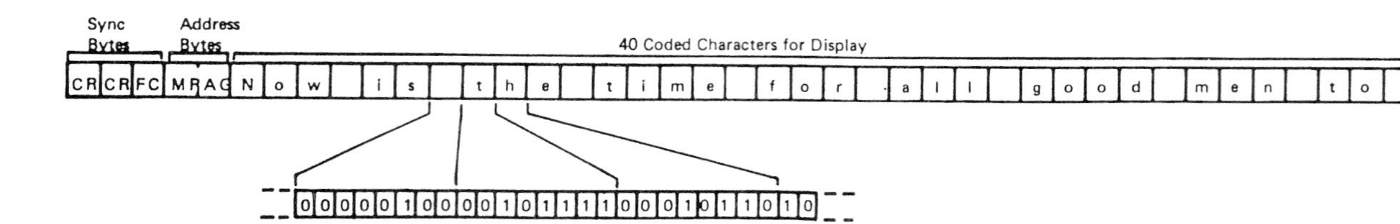

Figure 2.6(b) Row format (rows 1–23)

As each row of data for a particular page has the magazine and page number as its address they can be transmitted in any order and mixed with rows from other magazines. Rows containing no information need not be transmitted (rather than transmitting a row of 'space' characters). The technique of transmitting only these rows which contain information is called 'row adaptive' transmission. This feature is particularly useful when subtitles are transmitted, as these normally only contain one, two or possibly three rows of information; and also for news flashes, which again normally contain only a few rows of information. When the complete page has been transmitted the following page header row is used as an indication to the decoder that a complete page has been received.

In 525-line systems the video bandwidth is less than that for 625-line systems, while the active line period is approximately the same (52 microseconds). The data rate is therefore reduced by approximately the ratio of 525/625. This means that only 32 characters per row can be transmitted in one television line period, but 40 characters are required to be displayed. In the fixed format system, the 'gearing' technique is used to overcome this limitation [12]. The last eight characters of each of the previous four rows are transmitted as a separate complete row. Each page of 24 rows therefore requires 30 data lines. As the field frequency of 525-line system is 60 Hz, compared to 50 Hz for 625-line systems, the overall transmission rate for the system is similar.

Character control codes (control characters)

Character control codes are required to augment the display of normal alphanumeric display characters in order to create special effects. These include colour changes, flashing, double-height or double-width characters, and graphic symbols. Character control codes use the same 7-bit format as the characters and are often referred to as 'control characters'. In fixed-format systems the control codes are normally inserted in front of the block of text to which they refer and the transmission of a control code occupies a character space. This does not normally impose limitations to an editor since the control codes are inserted in spaces at the start of sentences or prior to a graphics symbol. Control codes contained within the page in this way are referred to as serial or spacing attributes. In free-format systems a space in the text is not needed for insertion of control codes, that is, they are 'non-spacing' attributes.

Character code extension

The character code tables for use with fixed-formatted pages are limited to a font of 96 display characters. Certain languages, for example, Spanish and Arabic, require additional characters. Such characters are transmitted on additional data rows that do not correspond to one of the 24 display rows and are therefore not displayed. Initially these rows were referred to as 'ghost rows', but they are now known as 'data packets' and have been assigned 'packet numbers'. WST has also been further developed to include high resolution graphics and other facilities, as outlined on pages 20–23.

Error protection

One transmitted data line carries 40 8-bit teletext character codes for one display row, preceded by a 3-byte start sequence and a 2-byte row address. An 8-bit character byte comprises a 7-bit ASCII code and a parity bit, the logic value of which always results in an odd number of logic 1s in the byte. The 7-bit code can have a maximum of 128 binary combinations. The data bits are numbered bit 1 . . . bit 7, and a byte is transmitted with least significant bit first and parity bit 8 last. The 128-code look-up table is shown in Figure 2.7 and the teletext code for any one of the 96 alphanumeric or 64 graphic characters can be found by combining the binary values of bits 1 to 4 on the left with the binary value of bits 5 to 7 at the head of the column. The 32 codes in columns 0 and 1 are control codes which switch the significance of codes in the other columns between alphanumeric and graphics and give some attribute, such as a colour change, to succeeding characters. Control codes are usually spacing and non-displayed characters, i.e. they occupy a character position in a row but are 'displayed' as blank spaces. (For editorial purposes, it is very often convenient to display control codes as two-letter mnemonics and this facility is available on editing terminals; see Figures 3.2 and 3.4.)

All teletext rows start with two bytes of 'run-in'. This forms a train of consecutive '1' and '0' which gives the maximum possible number of data transitions for excitation of the clock recovery tuned circuit in the receiver decoder. This establishes bit synchronism in the decoder. Byte 3 is the framing code, on recognition of which the decoder character counter starts up, thus establishing byte synchronism. Bytes 4 and 5 are Hamming codes, each containing four packet address bits and four error protection

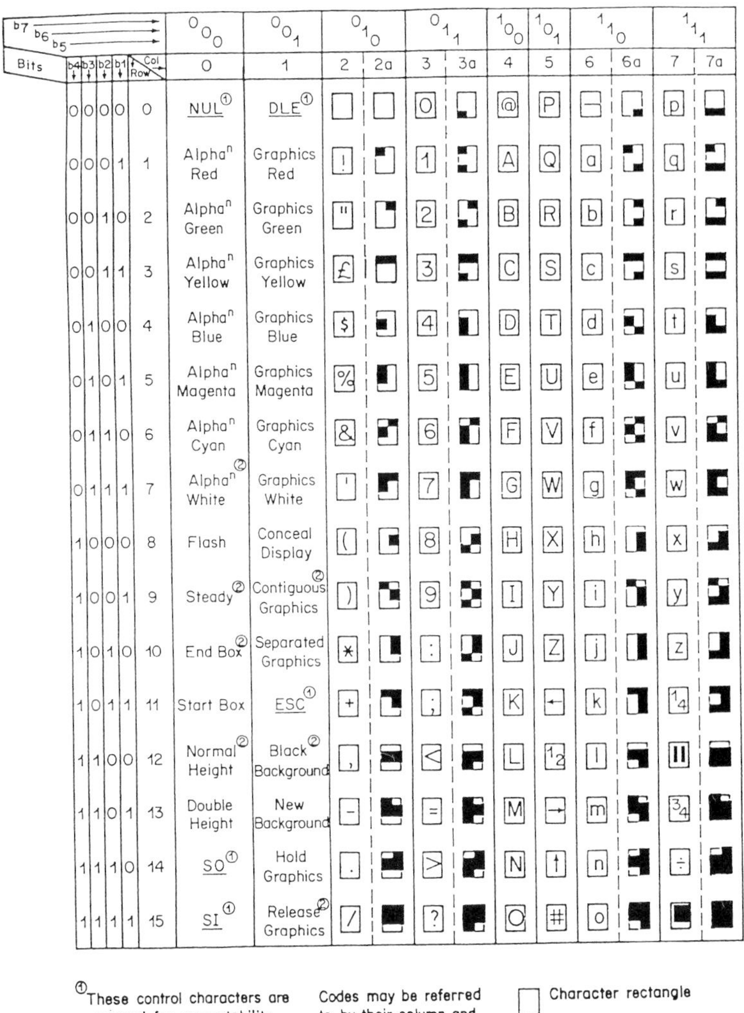

b7 b6 b5 →					0 0 0	0 0 1	0 1 0		0 1 1		1 0 0	1 0 1	1 1 0		1 1 1	
b4	b3	b2	b1	Row \ Col	0	1	2	2a	3	3a	4	5	6	6a	7	7a
0	0	0	0	0	NUL①	DLE①			0		@	P	—		p	
0	0	0	1	1	Alpha^n Red	Graphics Red	!		1		A	Q	a		q	
0	0	1	0	2	Alpha^n Green	Graphics Green	"		2		B	R	b		r	
0	0	1	1	3	Alpha^n Yellow	Graphics Yellow	£		3		C	S	c		s	
0	1	0	0	4	Alpha^n Blue	Graphics Blue	$		4		D	T	d		t	
0	1	0	1	5	Alpha^n Magenta	Graphics Magenta	%		5		E	U	e		u	
0	1	1	0	6	Alpha^n Cyan	Graphics Cyan	&		6		F	V	f		v	
0	1	1	1	7	Alpha^n White②	Graphics White	'		7		G	W	g		w	
1	0	0	0	8	Flash	Conceal Display	(		8		H	X	h		x	
1	0	0	1	9	Steady②	Contiguous Graphics②	)		9		I	Y	i		y	
1	0	1	0	10	End Box②	Separated Graphics	*		:		J	Z	j		z	
1	0	1	1	11	Start Box	ESC①	+		;		K	←	k		¼	
1	1	0	0	12	Normal Height②	Black Background②	,		<		L	½	l		‖	
1	1	0	1	13	Double Height	New Background	-		=		M	→	m		¾	
1	1	1	0	14	SO①	Hold Graphics	.		>		N	↑	n		÷	
1	1	1	1	15	SI①	Release Graphics②	/		?		O	#	o		■	

① These control characters are reserved for compatability with other data codes

② These control characters are presumed before each row begins

Codes may be referred to by their column and row e.g. 2/5 refers to %

□ Character rectangle

Black represents display colour

White represents background

Figure 2.7 Teletext 96-character code table

bits. The least-significant three bits of the 8-bit packet number is the binary-coded magazine number, identifying which of eight possible 100-page numbered magazines the row belongs. The most significant five bits are the binary-coded row address. Only 24 of the available 32 row-address codes are used for 'level-1' teletext (see pages 20–23). It is now common to express data row numbers as 'packet X/Y', where X is the magazine number and Y the row number. All data lines start with the five bytes as described, followed by 40 teletext character codes, i.e. 45 bytes per line.

The header row carries a further eight Hamming-protected address and control bytes, leaving only 32 character codes for display, as shown in Figure 2.8.

Bytes 6 and 7 are binary coded decimal (BCD) page number units and tens (00–99). When the combined magazine number and page number on the transmitted header agree with a user-requested magazine and page number in the decoder, the page is written into the decoder's page memory and displayed.

Hamming-coded bytes 8, 9, 10 and 11 provide for page number sub-codes which can either be related to time of day as indicated in Figure 2.8, or can be used as an extended address to increase the basic 800 decimal numbered pages to some 2.56 million, or even more using hexadecimal numbers. The extended address is in BCD, byte 8 is minutes (0–9), byte 9 is tens of minutes (0–7), byte 10 is hours (0–3) and byte 11 is tens of hours (0–2). Control bits C4, C5, C6 take up the spare address bits in bytes 9 and 11. Eight further control bits make up Hamming coded bytes 12 and 13. The control bit functions are listed in Figure 2.8. C12, C13, C14, shown as unallocated, are now used to switch decoders to display a basic character set of 83 characters, plus one of three sets of 13 national option characters. The editor decides which national option set is displayed by transmitting a particuar combination of control bits C12, C13, C14. Character generators with national options are available for various languages or combined in a single LSI circuit.

Hamming protection codes make correct reception of the addresses, that is magazine and page numbers, more probable; without these the following data cannot be displayed. The parity bit check for individual characters normally ensures that should an error be received the character will not be displayed. In the event of two errors in one byte, the decoder will display an incorrect character. Since the teletext magazine of pages is transmitted cyclicly, when the page is next received characters that were received with an error on the first acquisition almost certainly will be received and displayed correctly. Should a single error be received in a character that is already displayed correctly, then that character will not be changed. In free-format systems Hamming protection is essential for positional and control information as well as for the page address, which is a significant overhead.

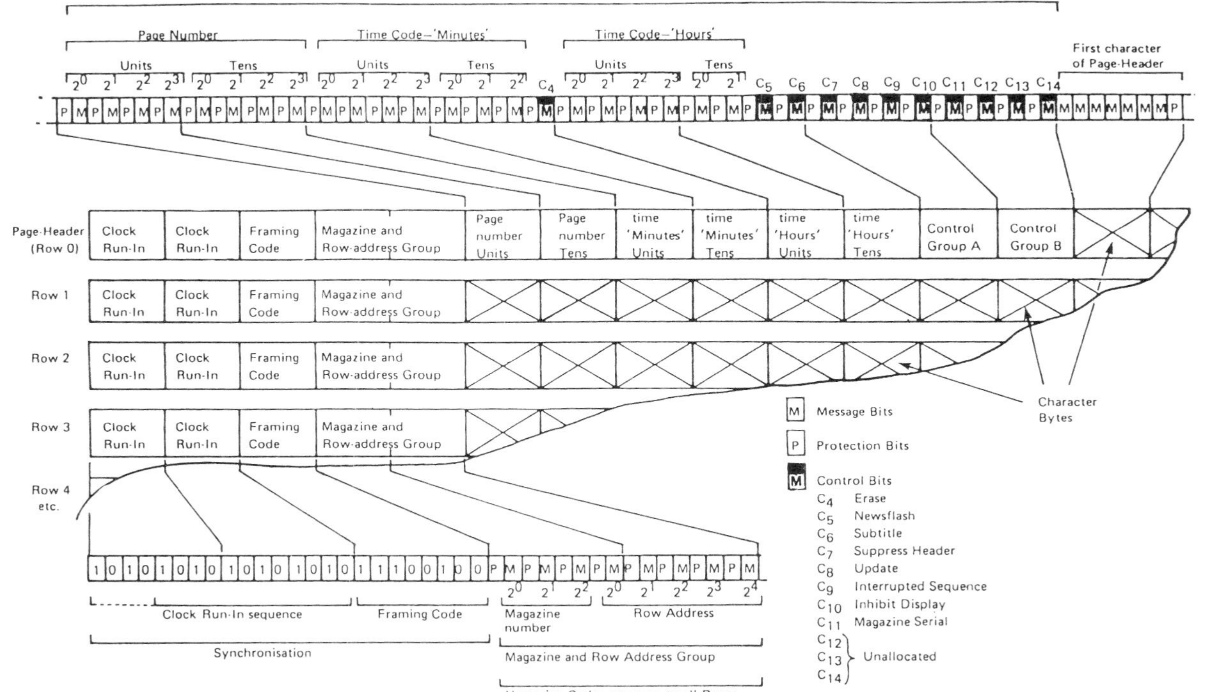

Figure 2.8 Synchronization and Hamming codes at start of page-header and row transmissions

Access time and data rate

Access time is the time taken to display a complete page after selection by the viewer and it is very desirable that this should be as short as possible. To obtain the maximum throughput of data in a teletext system it follows that the data rate must be as high as practical, bearing in mind the constraints of the television channel. For 625-line systems the video bandwidth is at least 5 MHz, (5.5 MHz for the UK television system I and 6 MHz for system D). When the teletext data standard was being evolved 625-line reception tests showed that the noise immunity of data transmitted over the television channel was not significantly less for data at a bit rate of just under 7 M bits per second than for data at a lower rate of 4.5 M bits per second. A bit rate of 6.9375 M bits per second was chosen for the 625-line systems. The bandwidth of 525-line television systems is normally 4.2 MHz and a data rate of 5.727272 M bits per second was therefore chosen. The teletext page is displayed in colour (on a colour receiver) but the teletext system and data signal is independent of the video colour standard (PAL, SECAM, or NTSC). The video signal simply acts as a carrier for the data, the decoder providing RGB output signals.

The data rate chosen for the 625-line WST enables one complete row of text to be carried in one television line period. The page is normally made up of 24 rows and it follows that approximately two pages are transmitted per second per data line used. When six data lines are used some 12 pages per second are transmitted. In this case the access time will be a maximum of about 16 seconds, with an average access time of about eight seconds, when 200 full pages are transmitted. Although the data rate is lower, 525-line fixed format systems have a similar access time as the field rate of 60 Hz is higher.

Higher levels of WST

As outlined in the introduction, WST has been further developed to cater for a wide range of future applications. These enhancements have been specified [13] as five levels, in a flexible manner so that the required features of any level can be implemented as required. However, current decoders exploit the various language capabilities of level 2 only. The main features of the various levels are outlined as follows.

Level 1

Ninety-six character font, double height, flashing, upper and lower case characters, mosaic graphics, eight colours, concealed characters, newsflash and subtitle pages.

Level 2

Level 2 introduces non-displayed rows, which are now called packets; the packet number corresponds to the row number. The teletext packet consists of 45 bytes. In every packet the preamble remains the same and is composed of a 2-byte run-in and a one-byte framing code which ensure bit- and byte-synchronization respectively. Immediately following this is the two-byte magazine and packet address group (MPAG), formerly known as the magazine and row address group. These packets contain control data and are transmitted before the teletext pages (Packets 0 to 23) so that the additional control data for the decoder arrives before the actual page. This speeds up processing and avoids any necessity to change the page after initial display.

Packets 0 to 23 inclusive are as level 1. They relate directly to the displayed page and the 40 bytes after the preamble are dedicated to the definition of characters and their display attributes.

Packet 24 is used to display the ‘Fastext’ prompts at the bottom of the screen, on a 25th row. These prompts take the form of four colour-coded keywords, e.g. News (Red), Sport (Green), Weather (Yellow) and Finance (Cyan). The colours match keys on the viewer’s remote control handset (see Chapter 5, pages 54–56).

Packet 25 consists of 40 display characters or attributes and overwrites the page header, row 0, on a decoder that supports this feature.

Packet 26 can support many modes but is primarily intended for extending a character set (typically from 96 to 128 different characters). This is achieved by using supplementary characters. The character overwrites the level-1 display at a particular row and column as defined within the packet. Fallback for level-1 decoders is defined by the editor, who ensures that a suitable character will be shown on level-1 decoders.

Whereas in level 1 each control character occupies a display space, in level 2 nonspacing control characters are carried in a packet. This can give more character spaces within the page for display use.

Changes to the character size in two dimensions are possible, give the ability to display characters both of double height and double width.

Packet 27 provides the 'Fastext' page links. Up to eight packet 27s are currently defined but a total of 16 are possible.

Packet 28 is allocated to define the display aspects of a particular page (apart from the header). A total of 16 packet 28s is possible; each packet contains a 'designation code' followed by typically 13 groups, each of three bytes of data. One mode of operation is to cause the decoder to select an extension character set or to switch from a Latin-based to a non-Latin one. A further use of packet 28 (with a different designation code) is to redefine pastel colours on a page basis.

Packet 29 is allocated to define the display aspects of an entire magazine.

Packet 8/30 is the broadcast service data packet. This packet is normally transmitted approximately once per second and carries a magazine and packet address group which is notionally equivalent to 'magazine 8 row 30', although it is neither a part of magazine 8, nor a row of any page. It can therefore be received only by using a special decoder.

Packet 31 is used for carrying general purpose data transmission services.

Level 3

Dynamically redefinable (down loaded) character sets (DRCS).
Down loading using pseudo-pages.
Linking from pages for display to pseudo-pages, for down loading character sets.
Pattern transfer units (PTU). Up to 96 PTUs can be down loaded using two pseudo-pages.

Level 4

Alphageometric displays.
Introductory pages.
Links to pseudo-pages.
Pseudo-pages for overwriting.
Pages for reformatable data.
Presentation layer syntax (geometric display).

Level 5

Alphaphotographic displays (still pictures).
Pseudo-pages carrying photographic data.

The data requirement for a level-1 page is 960 bytes and that for level 2 is limited to a maximum of 1920 bytes. A level-5 page may require 9600 bytes and hence the inclusion of such pages significantly increases access time.

3 Editing terminals and graphics digitizers

Editing terminals

The editing terminal used for creating teletext pages allows text, numerical information and graphics to be entered by the editor to create the pages simply without being concerned with any special teletext system requirements. The page being created is viewed on a colour monitor and the input keyboard is arranged in the same fashion as a normal typewriter but with additional keys for special effects such as colouring, double-height characters, etc. The editing terminal also includes a wide range of word processing features such as word wrap and tabulation to facilitate page creation.

Special software is therefore necessary in the terminal to simplify the editorial tasks. In some designs of teletext systems, much of the editorial software forms part of the main teletext system processor. This arrangement simplifies the editing terminals but continuous communication with the system is then essential and furthermore, the editorial tasks impose a very significant overhead on the system processor. This can result in a slow response when a number of terminals are in use and a high rate of output is required to insert data onto several FBI lines. The alternative arrangement, made possible by 16-bit microprocessors, is to increase the processing power of the terminal so that it functions as a self-contained work station. Such terminals only need to communicate with the teletext system processor when pages are being entered or changed or when system instructions are being given.

The keyboard layout of an advanced design of a stand-alone editing terminal is shown in Figure 3.1. The keys for colouring, tabulation, insert and delete are contained in the top row and many have appropriately coloured key cap marking. The text and numerical entry keys are arranged in the normal QWERTY layout. This area of the keyboard layout may be changed for different language requirements. The 12 cursor command keys are

Figure 3.1 Editing terminal keyboard showing layout of special keys

grouped in a separate area, as are those for the mosaic graphics cell, the editing grid and the page swap keys. The transmitted page consists of 24 rows of 40 characters per row, but the first row, row 0, is reserved for the service name, page number, time and date. This information is generated by the system processor and is not controlled by the editors. The 24th row at the bottom of the editing screen does not form part of the transmitted page and is often used for editorial commands to the system, as described later in this section. A typical editing terminal display is shown in Figure 3.2.

Figure 3.2 Editing terminal display showing command row (row 25)

The basic character font used in teletext has a capacity for 96 character shapes but many languages require a larger number to produce an acceptable display. For example, Spanish requires 128, and 131 are required for the Czechoslovakian language. The basic code table is used for the 96 most used characters and the additional ones are sent as a page extension packet. Packet 26 has been allocated for the purpose. The complete sets of upper case character shapes used for the Spanish and Czechoslovakian languages are shown in Figure 3.3 The terminal software

SPANISH

BASE 96 CHARACTERS						EXTENDED CHARACTERS	
	0	¡	P	¿	p	ö	À
!	1	A	Q	a	q	ï	·
"	2	B	R	b	r	â	ê
ç	3	C	S	c	s	ô	ã
$	4	D	T	d	t	õ	Ü
%	5	E	U	e	u	í	ä
&	6	F	V	f	v	ë	î
'	7	G	W	g	w	û	Á
(	8	H	X	h	x	Ñ	Õ
)	9	I	Y	i	y	Ã	Ç
*	:	J	Z	j	z	↑	E
+	;	K	á	k	ü	É	→
,	<	L	é	l	ñ	ª	º
-	=	M	í	m	è	ó	#
.	>	N	ó	n	à	ò	Ï
/	?	O	ú	o	■	ú	ò

CZECHOSLOVAKIAN

BASE 96 CHARACTERS						EXTENDED CHARACTERS		
	0	č	P	é	p	ó	Ň	
!	1	A	Q	a	q	ŕ	Ř	ď
"	2	B	R	b	r	Ǐ	Š	Í
#	3	C	S	c	s	ô	Ť	Ĺ
ů	4	D	T	d	t	ä	Ž	
%	5	E	U	e	u	Ó	É	
&	6	F	V	f	v	ß	Ô	
'	7	G	W	g	w	Ú	Ů	
(	8	H	X	h	x	§	ŧ	
)	9	I	Y	i	y	°	Ä	
*	:	J	Z	j	z	Á	ö	
+	;	K	ť	k	á	ň	ü	
,	<	L	ž	l	ě	Č	Ö	
-	=	M	ý	m	ú	Ď	Ü	
.	>	N	í	n	š	Ě	Ý	
/	?	O	ř	o	■	Ľ	Ŕ	

Figure 3.3 Extended character sets

automatically generates and sends the extension packet, with the page information, to the teletext system.

Latin-based languages are written left-to-right but languages using cursive script are often read right-to-left for text and left-to-right for numerals. Furthermore, in Arabic, the character shapes depend on the position in the text. There are isolated, medial, initial and final forms. A total of 120 characters are required to cover the various forms to provide an acceptable display. The Arabic terminal therefore uses a font of 120 characters and special software that automatically changes the shape according to the location of the letter in the text. Normal page display codes are used for 96 of the characters, the extra 24 being sent in the additional packet 26.

Special control codes are required by the decoder in the receiver to cause colour changes, double-height characters, flashing of words etc. These codes are produced by the terminal software, the required code being generated when the appropriate key is pressed. For example, if a word is required to be in red, then the red control key is pressed immediately before the word and the text is then automatically displayed in red until a colour change key is again pressed. The appropriate colour control code is generated by the terminal in the space preceding the word. Such control codes are usually displayed in receivers as blank spaces but an editing terminal has a facility to display a two-letter mnemonic to represent the code, if required, to assist the editor. The mnemonic chosen by the terminal designer usually represents the control code action, e.g. GG = Graphics Green; HG = Hold Graphics etc. The page shown in Figure 3.4 has the control codes displayed and is the same page as that shown in Figure 3.2. The page on the editing terminal is normally displayed as far as possible in the same way as it will be seen on a domestic receiver.

The editing terminal is normally able to store several pages in RAM, typically six or seven. This storage facility enables an editor to review a series of continuation pages off-line from the teletext system very rapidly. Graphics or text can also be copied from page to page. The area of text or graphics to be copied is marked using the cursor command controls. The area so marked is identified by a dotted line or background colour change. This area can then be moved on the page, duplicated in another part of the page or copied onto a separate page. The editorial term for this technique is 'cut and paste'.

Mosaic graphics are created using a special block of keys. Each graphics character is made up from a 2 × 3 cell format, so that 64 different shapes are available. The required shape is obtained by

Figure 3.4 Editing terminal display showing control characters

pressing the appropriate keys and repeating as necessary. When a graphics key is used in conjunction with the column mode key, vertical lines of graphics are automatically produced. All the necessary teletext control codes for colour and graphics shape are generated automatically by the terminal software without any editorial involvement. To assist the editor in laying out a page, a 23 × 40 grid can be displayed on the editing monitor. However, to produce a complex graphics 'picture', such as a map, using the keyboard only, is a very time consuming task and such pictures are normally produced using a graphics digitizer.

The functional diagram of an editing terminal is shown in Figure 3.5. The keyboard contains a processor together with the appropriate software that is associated with the various key functions. Different keyboard layouts may be used for different languages and the appropriate software would be contained in the keyboard processor. The keyboard is linked to the electronic unit by an RS232 data link operating at 9600 baud. The editorial processor would normally have the resident software contained in EPROM to avoid the use of mechanical disc drives and floppy discs being involved with the operation of the software. The RAM

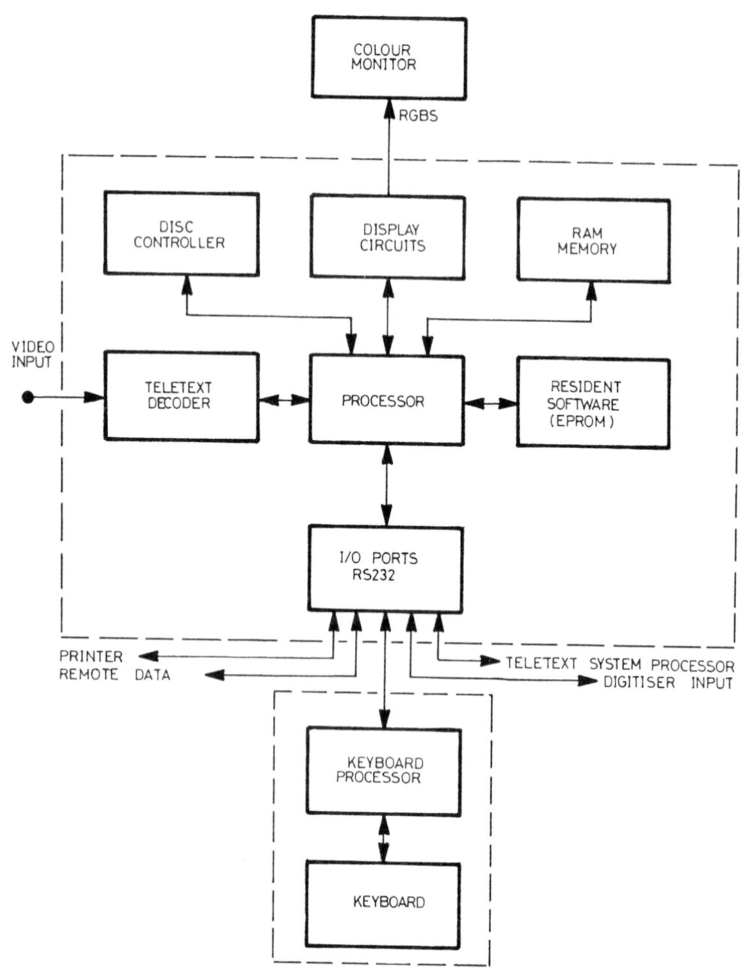

Figure 3.5 Functional diagram of an editing terminal

memory used for storing teletext pages, and other data used during the editing processes, is battery backed to prevent the data being lost in the event of a power failure. The processor also drives the display circuits which produce the RGB and sync signal for the colour monitor. The EUROM LSI circuit, widely used in television receiver decoders, is often used for the display circuit function to help ensure that the appearance of the page being created by the editor corresponds to that on a typical receiver.

The disc controller and associated disc drive is an optional feature that is often fitted to editing terminals. An editor can then produce a large number of pages off-line and stored on floppy disc, in advance of a sequence of events (as in a sports programme, for example).

The inclusion of a teletext decoder in the editing terminal is also an option. This allows an editor to capture a page 'off air' for checking purposes; alternatively, pages from another channel may be used to verify information, or possibly as an alternative data source.

Additional I/O ports are normally provided on an editing terminal. A printer can then be used to make records of any pages or information that is used. A remote data input would allow an editor to receive information on particular topics, such as financial information, from remote data sources. A side-by-side display mode is available on some terminals so that the external data can be displayed, in addition to the teletext page, to simplify editing. The digitizer input, previously mentioned, enables a digitized teletext page of graphics to be fed directly to the terminal. There is also an RS232 feed to the teletext system processor. Terminals are normally connected directly to the system, but modems are used when the editing terminal is located at some considerable distance. An example is an editing terminal located at an airport or sports event.

Communication and commands

The editor communicates with the teletext system processor and provides instructions to it by entering commands onto the command row, row 24 (Figure 3.2). This row does not form part of the transmitted page but is simply used to display the commands to the editor or to enable the system to respond. Depressing the command key immediately moves the cursor to the start of row 24 ready for the editor to type in the instruction.

After a page has been created to the editor's satisfaction, an instruction would be typed in on row 24 to cause the terminal to transfer the page from the terminal store to the teletext system. The command row contains the page number and any other instructions to the teletext system concerning that particular page. For example, if the page is part of a rotating or extended page set, then the display duration will have to be entered by the editor. Other system commands enable an editor to examine the magazine structure of the pages being transmitted and to 'enable' or 'disenable' pages from the transmission as necessary. Pages can also be updated or modified off-line from the system.

On editing terminals that are equipped with a teletext decoder, the instructions concerning which magazine and page should be captured are entered on the command row. Some commands (in general, destructive commands) must be entered twice to prevent an editor inadvertently erasing pages. In general, any executive commands to the teletext system processor which affect the transmission of pages have safeguards to prevent unwanted changes. To prevent unauthorized editing when a number of editing terminals are used in conjunction with the system, 'barring' may be included, so that only certain terminals can communicate with particular pages.

Communication to the teletext system processor is normally an RS232 feed at 9600 baud. This is a suitable speed for passing data to the processor without making the serial nature of the communication obvious to the editor. The page data structure of simple pages is sent to the processor in a serpentine form: reading from top left to the right and then starting again at the next row, and so on. Each row starts with a prefix code (row number) and has a CRC checkword following the 40 bytes of data. This enables the processor to determine the location on the page and to check the validity of the data.

Pages that have non-displayed rows or packets associated with them, for different language character sets or Fastext, have these rows sent first, normally in the order they are to be transmitted. A special prefix is always used on the last row of the page to indicate page end. This facility is used for row adaptive transmission, which is normally only used for subtitles and news flashes. Update information can also be sent in this manner to specific pages, which avoids sending a complete page. Pages at this stage do not contain the full teletext header (row 0), as this is added by the teletext system processor as previously mentioned. A normal teletext decoder cannot be used to decode pages for viewing at this stage, but the decoding process is undertaken in the editing terminal with special software.

Graphics digitizers

Graphics digitizers (sometimes called graphics cameras) are used to produce complex 'pictures' such as maps or pages for educational type programmes. The graphics digitizer is able to create the teletext page directly from a picture or drawing. The functional diagram of a mosaic graphics digitizer is shown in Figure 3.6. The picture to be digitized is positioned under a video camera, which is normally fitted with a zoom lens and mounted so that the distance between the camera lens and the picture can be varied. A source of controlled lighting is also provided. The video signal from the camera is clamped to hold the black level constant at the input to a slicing circuit; the slicing level is variable by the editor. The slicing circuit produces an output which is 'high' for components of the video signal which are positive with respect to the slicing level, and 'low' for those below. A monitor is provided to enable the editor to check the picture from the camera in the preview mode and can be switched to view the two-level or digitized video signal. This allows the editor to adjust the slicing level to produce the optimum picture.

The final teletext mosaic graphics page consists of 80 horizontal by 72 vertical cells. The next step is to convert the digitized video signal into a signal which is divided into this cell matrix. The 625-line television picture has approximately 290 active lines per field. The graphics digitizer uses 288 of them so that each cell corresponds to a row embracing four television lines. The digitized video signal is fed to a sampling gate which effectively chops the row into 80 discrete segments, each segment corresponding to one cell. The output from this circuit is fed to three memory circuits which are enabled sequentially by the timing generator. The stored outputs of these, together with a direct output, are fed to the graphics cell generator, which uses majority logic to determine whether the average of the four inputs is high or low (i.e. white or black).

Each graphics symbol in the final teletext display consists of a two-by-three matrix and there are 64 different combinations, depending on which of the cells in the matrix are black or white. Each of these combinations is designated by an ASCII code (see Figure 5.1). The state of each cell in each graphics character is registered and the state of each cell is fed from the register to the code generator to produce the appropriate ASCII code. Each character code is entered into the page store. The output signal is a serial bit stream of ASCII characters which is fed to an unsolicited input of the editing terminal. The editor is then able to use the terminal to make fine adjustments to the graphic shapes using the keyboard, and add the appropriate colour information together with alphanumeric script to complete the page. A graphics decoder is also fitted in the digitizer so that the editor is able to view the final teletext graphics page on the monitor before sending it to the editing terminal.

It is not normal practice to use a graphics digitizer with each editing terminal in an editing suite, but to switch the digitizer

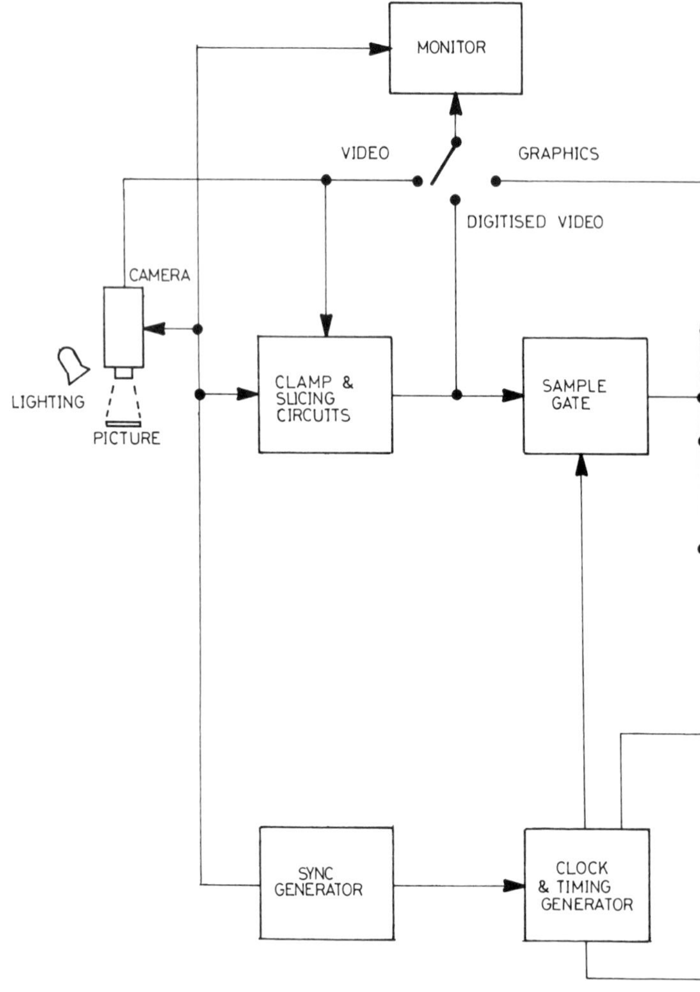

Figure 3.6 Functional diagram of a mosaic graphics digitizer

between terminals. Alternatively graphics pictures can be stored on a floppy disc before being transferred to the editing terminal for the final page preparation. The graphics digitizer is able to operate in real time so that the editor can move the picture and adjust the camera field of view to optimize the final teletext presentation on the monitor, using the internal decoding facility.

High-resolution graphics needed for level 3 require the picture to be analysed into the much finer resolution of 480 × 250 pixels.

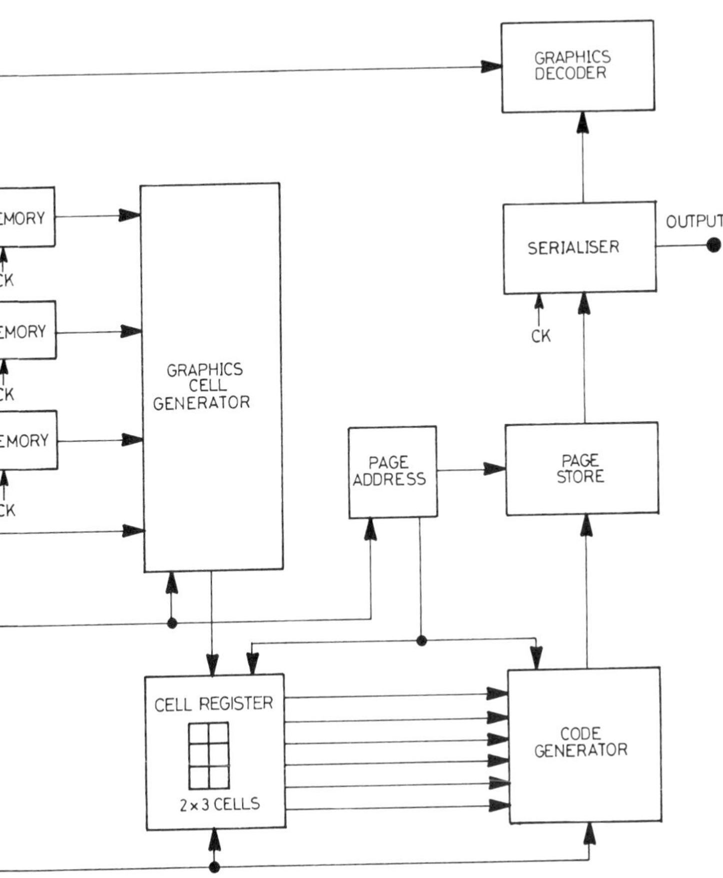

The video signal from the camera is clamped and sliced to produce a digitized or two-level video signal. This signal is then sampled on a one-pixel-wide sample basis. The television lines are sampled progressively, the sample position moving to the right by one sample period each field, as illustrated in Figure 3.7. The line samples are effectively about 64 microseconds apart. When the complete picture has been sampled and stored in memory it is then processed by the software.

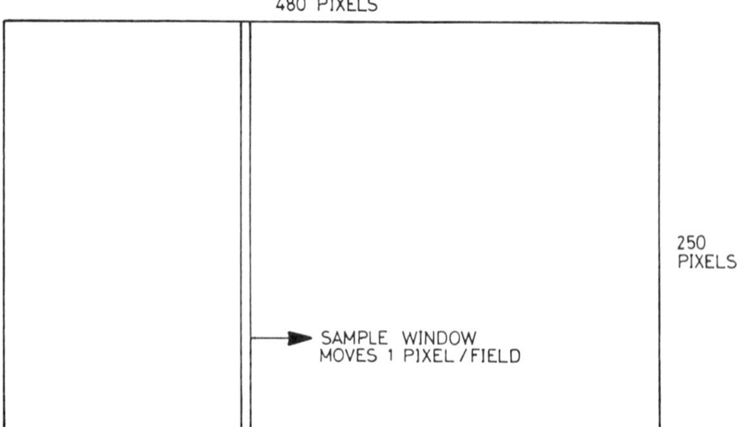

Figure 3.7 Level-3 sample structure

The picture being digitized is stationary and therefore a monochrome camera can be used in conjunction with coloured filters to produce the three digitized images of the RGB video signal. This enables the software to process the three digitized images directly into the teletext format complete with colour information. A complexity that occurs with level 3 is that the complete image cannot be digitized for transmission because the standard allows only a certain number of DRCS characters per page and therefore only part of the full screen image can be processed. Software algorithms are used to make the best fit with the smoothed graphic shapes and so reduce the number of DRCS characters required to construct the final page. The output circuit of the digitizer is arranged to display the graphics page on a colour monitor. The monitor can also be switched to display the monochrome video signal or the two-level digitized video signal so that the editor's optimization of slice level, position and orientation of the image is facilitated. When the editor is satisfied, the page will be finally processed and transferred to the editing terminal via an RS232 link, to enable alphanumeric information to be added as required.

Editing suites

Teletext is primarily designed for the rapid dissemination of information, and the main requirement of the editor is an adequate supply of editorial material. The editing suite will therefore require agency news feeds and be linked to a range of external information sources. The layout of an editorial suite is therefore not restricted by any technical requirement other than data communications.

External information sources can be linked directly to a teletext system, without editorial involvement, provided the appropriate page format and communication protocol is used so that the remote source emulates an editing terminal. When reformatting of the information is required, the necessary software is usually incorporated in the data management facility of the teletext system.

4 Teletext data management and transmission

The teletext pages from the editing terminals are fed to the main teletext system, which consists of a processor with resident software, the page store with adequate capacity for all the pages that will be used, the teletext data formatter, data pulse shaper and inserter. An archival store may also be provided together with a printer. The essential elements of a typical system are shown in Figure 4.1. The resident system software contains a wide range of system functions which the editors use to produce the magazines of pages to suit particular requirements. These include the more frequently transmitted pages, enabling a page to be transmitted out of sequence; this facility is sometimes used for index pages. A page may be transmitted, say, every 20 pages to provide rapid access for the viewer. The rotating, or extended, page sets are also produced by the system software, so that four, five or more pages may be given the same page number and be transmitted sequentially with a time interval of typically 20 seconds between each. This facility allows a teletext page to have several associated pages. Such instructions to the system processor are sent as commands from the editing terminals and are associated with the particular page (or pages) to which they refer. The system processor will have a number of entry ports for editing terminals and external data feeds. The system software will therefore contain a command structure designed for the particular requirements for processing the pages from memory into the required sequences for the output magazines together with data management software which is associated with the input terminals and other data inputs.

A hard disc can be used for both page storage and page archival purposes. However, the use of a hard disc for the page store limits the output capability of the system in terms of data output rate, due to the time delays in accessing the hard disc. Also system reliability is impaired because the mechanical disc drives are in continuous use. Large capacity RAM memories are now available at sufficiently low prices to allow them to be used for the page

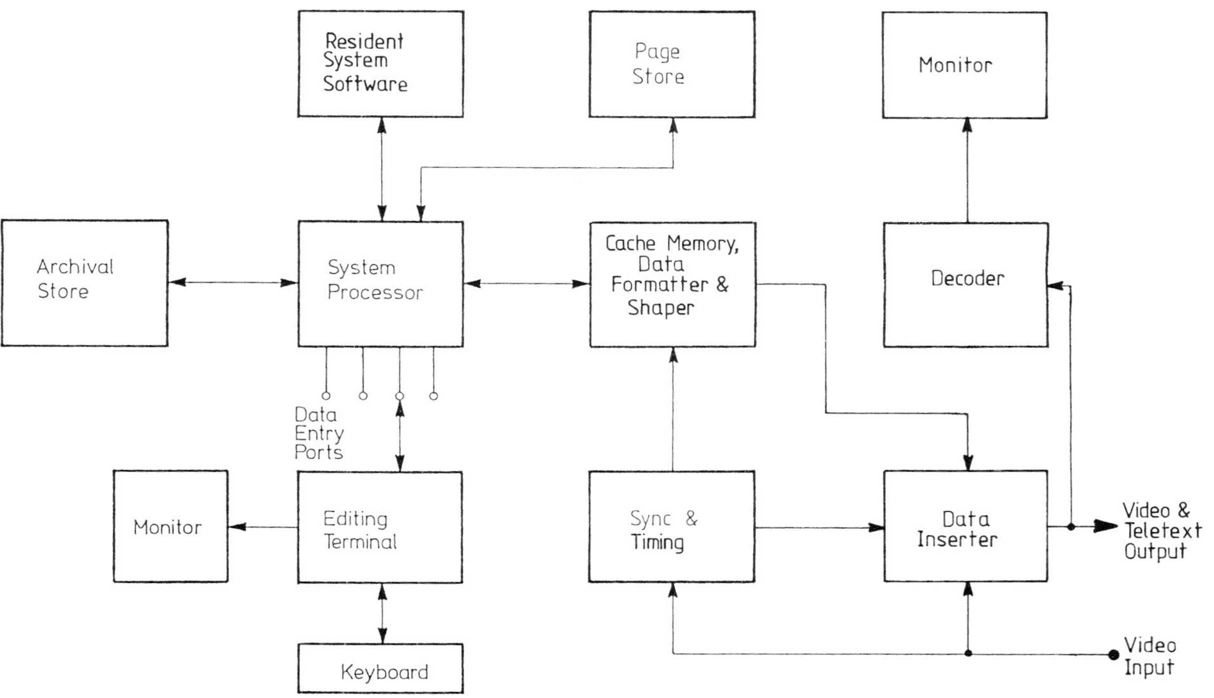

Figure 4.1 Single channel teletext system

stores, even on large systems. Such memories are battery backed so that in the event of a power failure, or when the system is switched off, the content of the memory is maintained. It is also becoming increasingly common for the resident software to be contained in EPROM rather than on a hard disc. This again improves system reliability since the software is completely independent of mechanical disc drives. On systems in which the operating software is contained on a disc, updates can be made or design 'bugs' cleared using a floppy disc or a modem link. Updating the software contained in EPROM requires the PROM to be changed, but the improved reliability is often considered worthwhile. When a hard disc is used for the archival page store a tape streamer is often used for backup purposes.

The system processor feeds the pages, in the required sequence, to a cache memory for the data formatting function. The header row (row 0), which contains the static information, the real-time clock and the data run-in sequence, is added, so that the page is formatted ready for transmission.

The formatted data is serialized and passed through a fast digital logic circuit to ensure that the data pulses have identical rise and fall times. This data is then fed through a special band shaping filter whose characteristics are carefully chosen so that the output signal meets the required broadcast specification. This signal is then fed to the data inserter for insertion into the video output of the studio. Synchronizing and timing signals are generated from the input video signal to control both the data formatter and the data inserter. Essential features of these circuits are that they time the data correctly, and insert it, once only, onto the selected FBI lines. Under no circumstances must the data ever be inserted during the video period, even when the sync pulses are disturbed by the transients that may occur in television networks (due, for example, to genlocking). The cross-talk from the fast logic circuits into the video circuit must be at a level of less than −70 dB with respect to the video signal. The data inserter must also be fitted with a physical by-pass relay, to operate in the event of a malfunction or power failure.

Backup facilities are also necessary for the teletext system so that a service can be maintained in the event of a malfunction in one of the units. In such a case it may be permissible to operate with a reduced service, for example, by limiting the number of editing terminals in use and having only a single magazine output. If no breaks can be allowed in the service, then a complete backup for the system to cover all possible modes of failure is necessary. A data monitor checks the output data leaving the system and should it detect any form of failure it will cause the standby units to be switched into circuit automatically, and raise the necessary alarm.

In many cases a teletext system may be required to feed two independent television channels with different information, together with some common information. Such a system is likely to use 10 or 12 editing terminals together with data inputs from other sources so that the management of the data inputs is itself a considerable task. In such systems the tasks of input data management and teletext system management are normally separated, with individual processors and software. Separation of teletext systems into smaller units with individual processors, each with relevant software, enables greater flexibility in design; and reliability can also be improved.

A dual-channel system with full backup is shown in Figure 4.2. The various functions are shown as separate units for clarity but in practice some of these might be combined. For example, one transmission processor might control the memory used for the two page stores for one channel, and also provide the two teletext outputs to the data detector. The various editing terminal and data inputs are fed to a data port switcher which feeds them to one of two identical data management systems. These systems operate in parallel so that all the information in one is duplicated in the other. When a large number of editing terminals are in use it is often necessary to restrict the access which they may have to the system, by the use of passwords.

The different magazines of pages are prepared by the data management software from the instructions provided by the editors. Magazines can also be prepared off-line, in advance, and can be ‘enabled’ at a specific time for transmission, entirely under machine control. A page which is required to be inserted into different magazines need be created only once; the data management software will insert it into magazines specified by the editor. The now familiar ‘World Time’ page will also be regularly updated automatically.

To simplify editorial tasks, new commands can be produced that cause a sequence of actions to take place from a single entry. Such commands are called ‘macros’. A typical example might be that a news page is entered into its own location and in addition be entered into other magazines and the news carousel. Archived pages that are not in current use but that may be needed at a future date can also be stored on the hard disc. As a hard disc can ‘crash’ and lose the data, the discs are normally backed up by a tape streamer which takes data from the hard disc and stores it on the tape cassette.

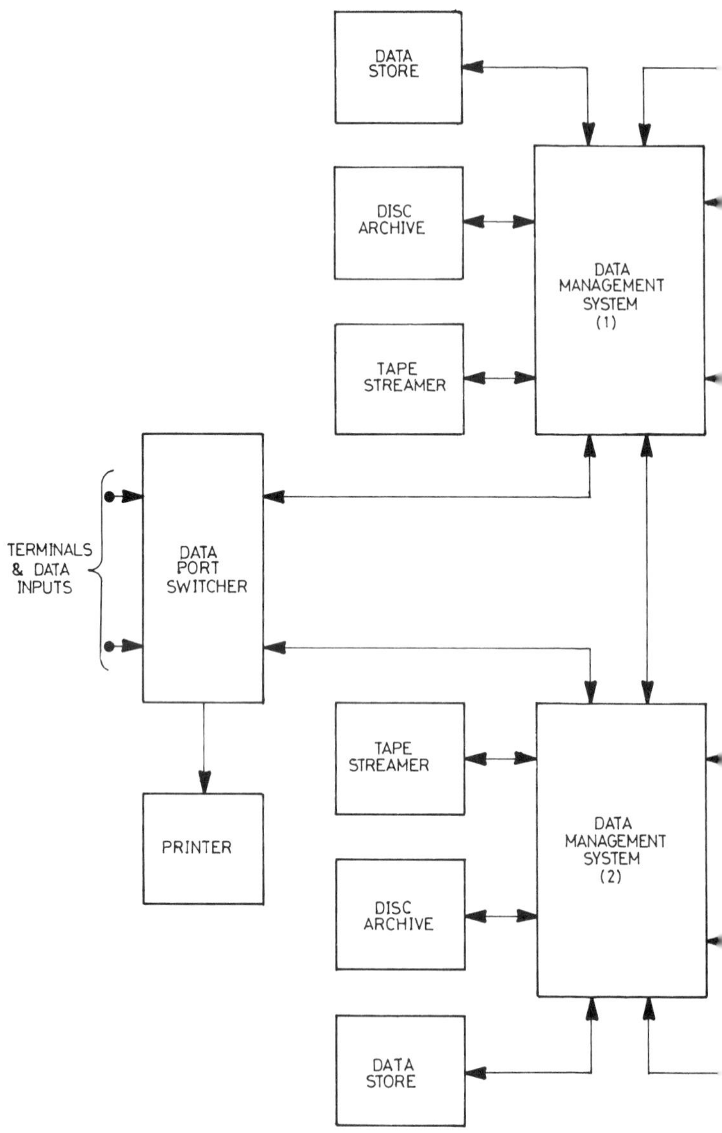

Figure 4.2 Dual channel system with full backup

External data feeds are also used by teletext systems so that special pages or data packets can be produced without any editorial action – for example, pages of financial information and

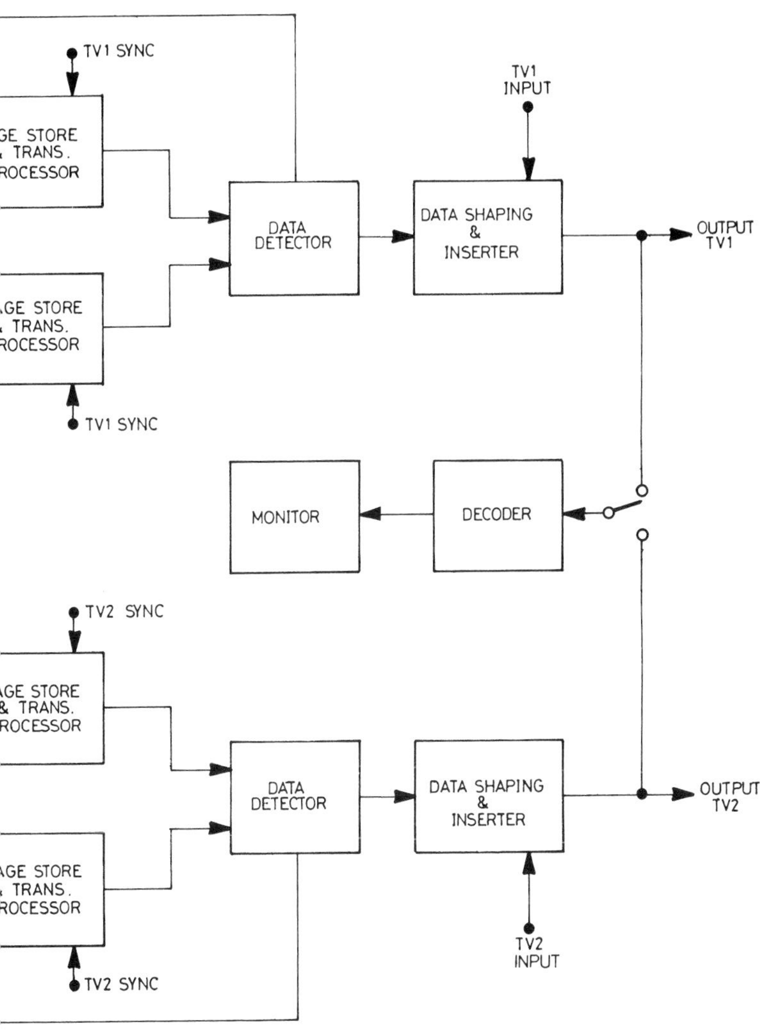

racing results and Packet 8/30 for the control of domestic video recorders. The data from these sources is fed direct to the system input ports using modem links for external information sources.

The data management software is programmed so that these unsolicited inputs are correctly page-formatted, the appropriate command instructions added (page numbers etc.), and the data entered into the correct magazine. Such inputs to the system would be archived, as would all the inputs from the various editing terminals, thus keeping a complete record of all data transactions. As many broadcasting authorities have to keep records of what is transmitted, there is an automatic logging system to ensure that all transactions are placed on record for a given length of time. A printer is normally used for transaction logging purposes. The time-code input from the station clock is also an automatic data feed to the system, which keeps the real time clock on the transmitted pages in step.

When several editing terminals are in use a 'mail box' facility can be provided to enable messages to be sent between terminals via the system. The communication channel is normally opened simply by specifying the source and designation names.

Each data management system drives a pair of cache memories and associated data formatters. One pair is cross-linked and the outputs combined in the data detector of each channel. The data detector monitors the teletext data activity and if this becomes static it switches the standby system into operation and raises an alarm. The serial data is shaped and inserted onto the FBI of the appropriate television signal. The teletext signal from a transmitter is normally continuously monitored to check the actual station output.

When subtitles are being transmitted monitoring the output is most important as the timing of the subtitle is critical. Preparation of subtitles (see Chapter 7) is normally an activity separate from the provision of the main teletext service. The subtitle output is fed to the teletext system as an unsolicited data input but the port must be given priority as the subtitle must be transmitted immediately it is received. The page currently being transmitted must be interrupted by the subtitle page to minimize transmission delay. The subtitle page is transmitted with the suppressed header bit set so that the header is not displayed in the receiver, the subtitle itself consisting of only one or two data lines.

The teletext system is self-contained as all the data is independent of the television service until the formatted data is inserted onto the fbi lines of the video signal. The necessary synchronizing information can be derived from the inserter, which is normally in the studio feed to the transmitter. The teletext system and the associated editorial activity is technically independent of the television service and can be located to suit the broadcast authority.

5 Teletext decoders and 'in-vision' systems

Basic functions

The teletext decoder, when incorporated in a television receiver, consists of additional integrated circuits associated with the video processing section of the receiver. The decoder input circuit is normally fed with a video signal of some two volts amplitude derived from circuits associated with the video detector. The decoder output provides RGB and blanking signals of a similar amplitude which are fed into the low level RGB amplifiers of the receiver. Teletext decoders are incorporated only into receivers which have remote control. The control signals for the decoder are therefore derived from the remote control circuits [4e].

The major differences between the various teletext systems are seen in the transmission time, which is related to the amount of data required to be transmitted for a given page, and in the complexity of the receiver decoder. The differences in the decoders are principally concerned with the processing that needs to be carried out on the data stream before the information is in a suitable form to be displayed. However, the input and output requirements for the interface between the decoder and the receiver video circuits, and the functions of the input and output circuits of the decoder, are similar in all systems.

Input circuits

The composite video signal is applied to the data acquisition circuit, which is used to process and select the required teletext data for a particular page so that it can be written into memory. Teletext data is recognized by the framing and page address codes irrespective of the actual line in the FBI on which it occurs. The acquisition circuit also performs data slicing, and this function is probably the most critical as it determines the reliability of data reception in poor reception conditions. For optimum performance the slicing level should normally be at 50% of the data pulse height so that the effect of pulse distortion has minimum effect. Video distortion or co-channel interference may cause variations in black level during the line period, so the optimum slice level will vary during the data line. The acquisition circuit therefore must contain circuits to clamp the video signal and to set the data amplitude at the optimum slicing level, to produce the cleanest possible data signal. The timing functions and data clock, which is locked to the incoming data stream, are also generated by the acquisition circuit. A crystal oscillator normally forms part of this circuit function, the crystal operating at a frequency corresponding to the data rate or a multiple of the data rate. Error-correction circuits are also incorporated into the acquisition circuit. The Hamming-protected bytes are checked and those having only a single bit error are corrected.

Alphanumeric characters forming the page content are transmitted in 8-bit data groups formed from 7-bit character codes plus one (odd) parity bit. The parity bit check for individual characters normally ensures that should an error be received the character will not be displayed. If two errors are received in one byte the decoder will display an incorrect character. Since the teletext magazine of pages is transmitted cyclically, when the requested page is next received characters that were received with an error on the first acquisition will almost certainly be received correctly the second time round and therefore displayed correctly. Should a single error (or odd number of errors) be received for a character that is already displayed correctly, then that character will not be changed.

The viewer selects pages using the remote control handset and the data derived from the remote control system is applied to the data acquisition circuit. This control data will specify the page and magazine number of the page that is required to be displayed. A comparator system is incorporated into the circuit so that only requested data is processed by the error correction circuits. The corrected data is fed, to the memory, as parallel data.

Output circuits

The decoder output circuit contains a character generator ROM for converting the 7-bit character code into a dot matrix pattern form which, in its simplest form, is a 7 × 5 matrix. The ROM contains the complete set of symbols which may be displayed. A typical device would contain at least 96 symbols, which can be selected by means of the 7-bit ASCII code (American Standard Code for Information Interchange) applied to the input circuit. The code table for a basic 96-character ROM is shown in Figure 5.1 and it can be seen, for example, that the code for A is 1000001. Each character is formed from appropriate dots in the 7 × 5 dot matrix. The ROM is operated on a row scan system, which means that all the dot information in a horizontal row of a character is available simultaneously at the output. A second set of inputs,

Bits b7 b6 b5 →					0 0 0	0 0 1	0 1 0		0 1 1		1 0 0	1 0 1	1 1 0		1 1 1	
b4	b3	b2	b1	Row \ Col	0	1	2	2a	3	3a	4	5	6	6a	7	7a
0	0	0	0	0	NUL	DLE			0		@	P	–		p	
0	0	0	1	1	Alpha^{n} Red	Graphics Red	!		1		A	Q	a		q	
0	0	1	0	2	Alpha^{n} Green	Graphics Green	"		2		B	R	b		r	
0	0	1	1	3	Alpha^{n} Yellow	Graphics Yellow	£		3		C	S	c		s	
0	1	0	0	4	Alpha^{n} Blue	Graphics Blue	$		4		D	T	d		t	
0	1	0	1	5	Alpha^{n} Magenta	Graphics Magenta	%		5		E	U	e		u	
0	1	1	0	6	Alpha^{n} Cyan	Graphics Cyan	&		6		F	V	f		v	
0	1	1	1	7	Alpha^{n} White	Graphics White	'		7		G	W	g		w	
1	0	0	0	8	Flash	Conceal Display	(		8		H	X	h		x	
1	0	0	1	9	Steady	Contiguous Graphics	)		9		I	Y	i		y	
1	0	1	0	10	End Box	Separated Graphics	*		:		J	Z	j		z	
1	0	1	1	11	Start Box	ESC	+		;		K	←	k		¼	
1	1	0	0	12	Normal Height	Black Background	,		<		L	½	l		‖	
1	1	0	1	13	Double Height	New Background	-		=		M	→	m		¾	
1	1	1	0	14	SO	Hold Graphics	.		>		N	↑	n		÷	
1	1	1	1	15	SI	Release Graphics	/		?		O	#	o		■	

Figure 5.1 Code for 96-character ROM

known as row address input, provides the vertical element of the character by determining which of the seven rows is supplied to the output. The dot information is then fed out in a serial form at a rate determined by the write clock, typically about 6 MHz. This arrangement is illustrated in Figure 5.2. The output signal forms the video output, the character colour being determined by the colour control circuit which selects the appropriate combination of RGB output signals.

'Character rounding' can be used to improve the resolution of the characters and makes use of the fact that all the letters, numbers and symbols in normal use are made up of a series of lines. The character rounding circuit modifies the video signal from the character generator when a diagonal line is being generated so that a smoother display is obtained. The effect is

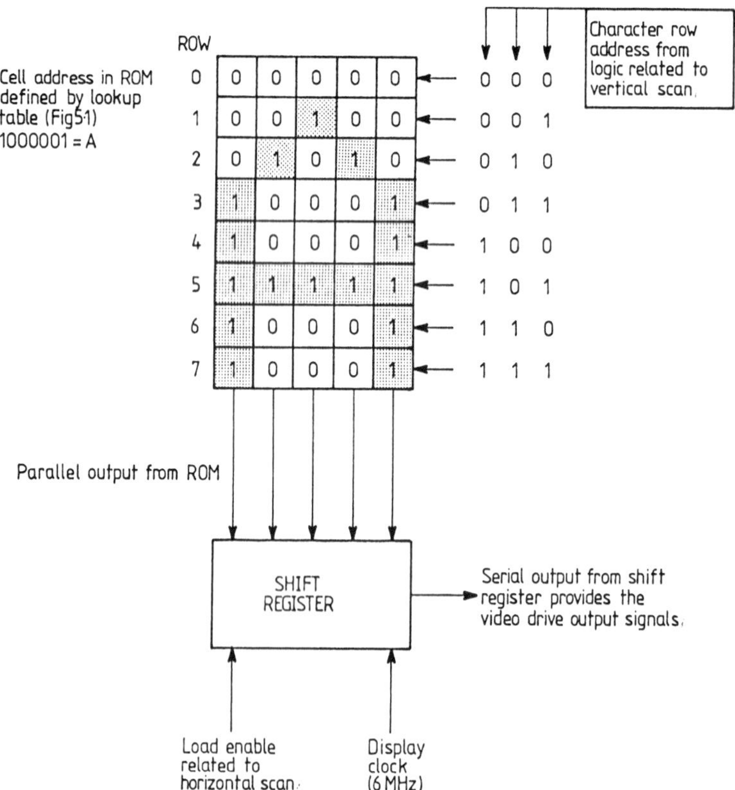

Figure 5.2 Operation of basic character generation ROM

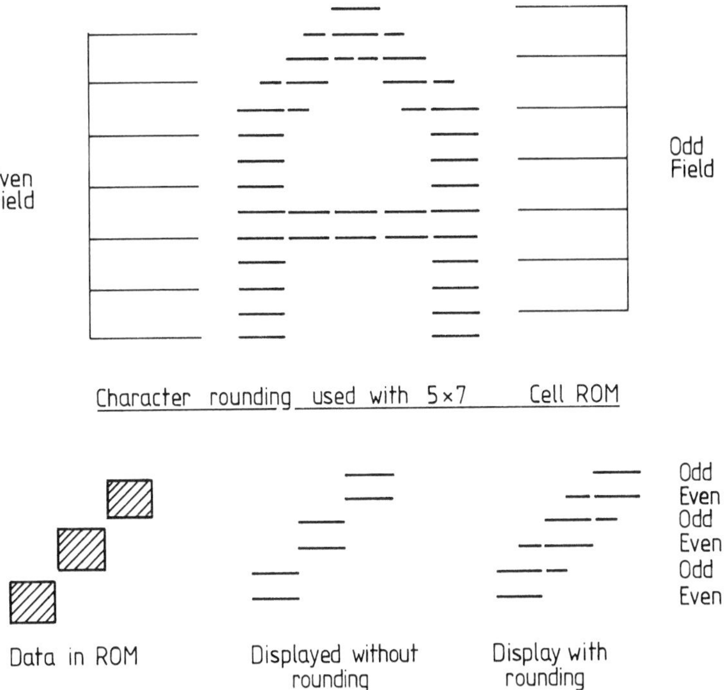

Figure 5.3 Typical character and location of half dots for character rounding

illustrated in Figure 5.3. No increase in the bandwidth of the video amplifer in the receiver is required for this improvement. In effect, the dot resolution of each character is doubled in both the horizontal and vertical directions. A teletext character generated using a ROM containing 7 × 5 dot matrix characters is displayed with a 14 × 10 dot resolution.

Display of alphanumeric data on an interlaced television raster gives rise to 'interline flicker', which can be objectionable. This effect is overcome by using a non-interlace (288-line) display for teletext pages. Whilst this effectively removes the interline flicker effects it also prevents the use of character rounding. Recently developed decoder circuits enable an interlaced or non-interlaced display to be switched as a design option. At the same time the dot matrix in the character generator ROM has been made 9 × 10 to improve the character resolution. Interlace must be used for subtitles and news flashes, which are inserted into the video picture.

Certain control functions are also performed in the decoder output circuit. These include the selection of graphics or alphanumerics and flashing of words or symbols. A blanking signal for use in the video circuits when a teletext page, or part of a page, has to be inserted into the video signal is also produced by the output circuit. This is required for news flashes and subtitles as these are normally displayed as boxes within the television picture. Timing signals are also fed to the output circuit to ensure that the RGB video and blanking signals are correctly timed with respect to the receiver display.

The ROM (or ROMS) contained in the output circuit must contain the complete character fonts required for various languages that are required to be displayed. Recently developed output circuits also contain an area of programmable ROM. Special character shapes may then be down loaded from the editing system to enable the receiver to display high-resolution graphics or the special characters required for certain languages such as Chinese. Such displays require more information than that contained in a normal teletext page of 960 characters. Larger page memories are therefore required in the decoder and the access time when such pages are being used becomes correspondingly longer (Chapter 2, page 20).

Free-format systems

Decoders for use on free-data format require two areas of memory, together with a processor and the appropriate resident software. The functional diagram of a basic decoder is shown in Figure 5.4. The data from the de-multiplexing circuit is fed into the acquisition memory. The processor reformats this data in accordance with the control and positional information to meet the display requirement and loads this processed data into the display memory. The parallel output of the display memory feeds the character generator ROM and output circuits.

Fixed-format system

Decoders for the World Standard Teletext (WST) fixed-data format systems require only a single memory and no processor as the data from the acquisition circuit is already in page format form and the memory therefore feeds directly to the character generator and output circuits. The functional diagram of a basic decoder is shown in Figure 5.5. This simplification of the decoder function

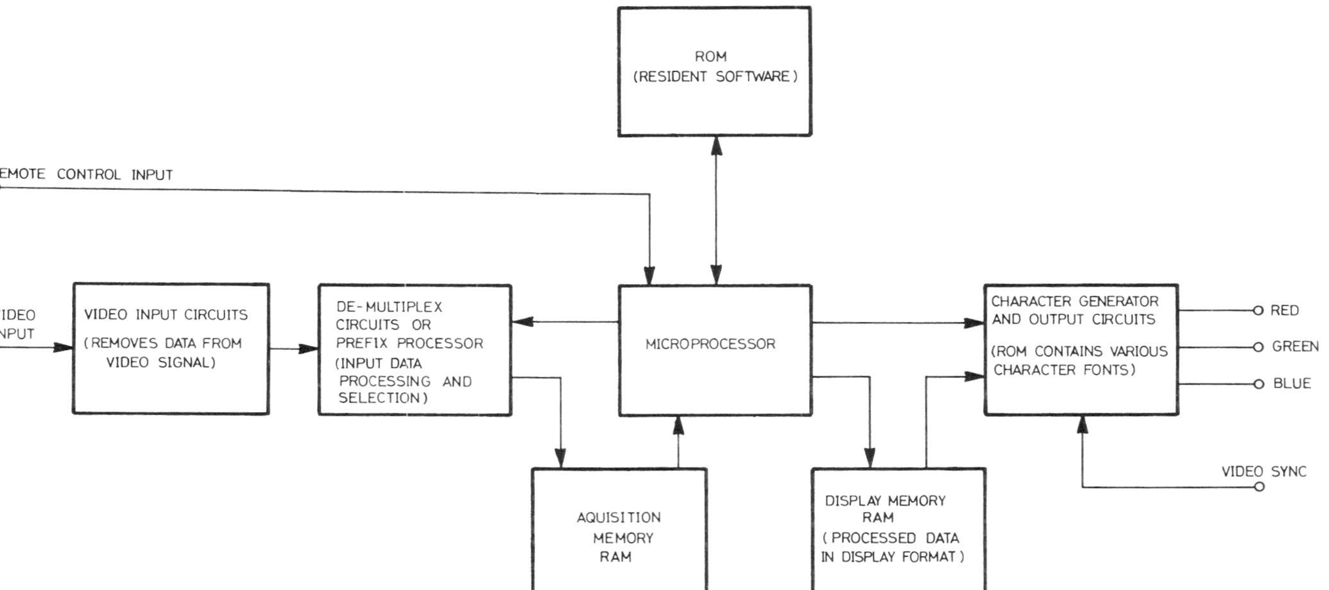

Figure 5.4 Functional diagram of free-format decoder

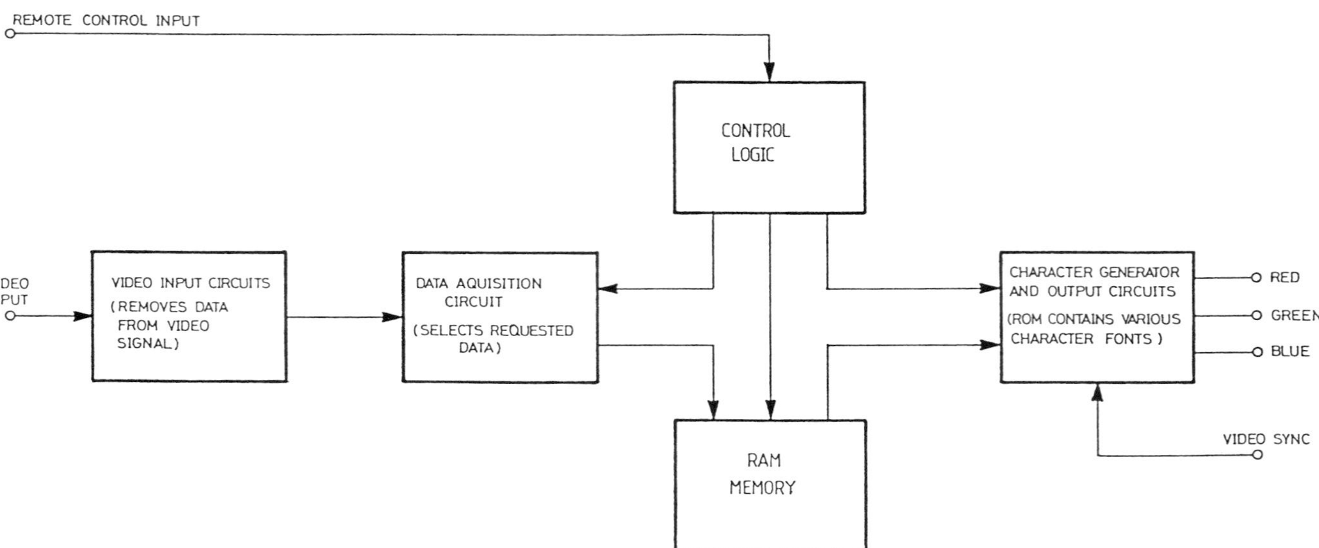

Figure 5.5 Functional diagram of fixed-format decoder

means that the fixed-format systems can display information virtually instantaneously, as it is received, and the absence of the processor means that the decoder circuits can be very simple. The higher levels of WST teletext which employ data packets require additional memory together with a processor and the necessary resident software.

Decoder performance

The performance of a teletext decoder, or of a complete teletext receiver, is normally judged by the errors in the displayed picture. The teletext data signal is digitally coded alphanumeric data and provided the '0' and '1' levels of the data stream are well separated the decoder is able to decode the data without errors.

Degradation of the received data will gradually reduce the separation between '0' and '1' levels until a point is reached when the decoder cannot make correct decisions and errors will start to be produced. The address information, that is the magazine and page number and, in the case of a free-format system, the necessary positional information, is protected by Hamming code. Hamming protection of magazine and page number ensures that the decoder will process only that data which corresponds to the requested page. When the page is first received any initial errors will show up as an incorrect character on the page of text. When the page is received for the second time the decoder will usually correct these display errors. Thus initial errors only show up when the page is first received. The teletext service can be considered at its limit when (typically) three or four errors are received when a page is first received. As the signal impairment increases the rate of errors increases rapidly, making the page unusable. The edge of the service area for teletext, unlike that for normal colour television reception, is relatively abrupt, the transition from good performance to unusable occurring quite rapidly, as is common for all digital systems.

The main signal degradation which causes errors to teletext reception is that due to reflections. These distort the pulse shape, causing confusion between '1' and '0' levels in the data stream. The teletext data signal is also degraded by noise and co-channel interference. Impulsive interference does not normally cause any problems unless it occurs during the period when the teletext data signal is being acquired by the decoder, that is during the FBI.

The accepted design target for receivers is to operate with an input signal having a decoding margin (Chapter 9, page 97) of about 25%, with only two or three errors occurring when the page is first acquired. Allowing for degradation in the tuner and IF circuits, the decoder would need to operate with an input signal having a decoding margin of about 20%.

Signal path distortion

To optimize teletext performance the distortion that can occur in the IF amplifiers and video demodulator must be minimized. The use of surface wave filters (SWAF) significantly improves the performance of IF amplifiers, compared to LC block filters.

The video demodulator is the main source of non-linear distortion. Simple diode demodulators, which respond to the envelope of the RF signal, introduce a high level of quadrature distortion. This seriously affects the data eye characteristics, causing a loss of eyeheight and symmetry (Chapter 9, page 93). The vestigual side band transmission requires a detector which responds only to the modulation component in phase with the carrier. Fully synchronous demodulators give the best eyeheight performance, but cost prevents their widespread use in domestic television receivers. However, 'quasi-synchronous' demodulators perform well with both television signals and teletext data.

Receiver de-tuning has more detrimental effect on teletext reception than that on the reception of the television video signal. A tuning accuracy of ±50 kHz is desirable and thus a high performance AFC system is necessary if digital tuning is not used.

To measure and assess the performance of teletext decoders accurately requires a source of signals with controlled levels of distortion, and special signal generators are produced for such tests (Chapter 9, page 99).

In domestic locations teletext performance can easily be marred by poor aerial installations and reflected signals. Experience has shown that when an aerial installation is adjusted for good teletext performance the resulting colour television picture is usually significantly improved. In general, good teletext reception is more difficult to achieve with transmissions in Band I because receiver aerials are much less directive than those used for the higher frequency bands, and at the same time electrical interference from motor cars can be more troublesome.

Multi-page decoders

Decoders capable of storing several teletext pages are now starting to be used in WST receivers. A processor is also incorporated in such decoders to control the acquisition and storage of required pages. Various strategies are being employed to reduce the access time and also to make the decoder more 'user-friendly'.

An arrangement that does not involve any editorial function is to incorporate in the decoder a memory with an eight-page capacity and to arrange for the capture of the seven pages subsequent to that requested by the viewer. This allows the viewer to step through the magazine with a virtually instantaneous display of the requested page.

An alternative arrangement that is now specified as a WST option [13] is to use additional information provided by the editor. The editor adds this information to the page in a form of a data packet, that is a row of data that is not displayed by the decoder but is used to instruct the decoder which additional pages it should capture. This instruction is based on the editor's anticipation of the viewer's requirements. The decoder is also made more 'user-friendly' by arranging that a command from a single key press on the viewer's handset selects both magazine and page number. Special additional information is displayed on the bottom row of the page to provide simple instructions for the viewer's use. For example, the additional instruction row might contain four topics such as sport, news, financial and travel information. Each of these topics would have a different colour background, and buttons on the handset would have corresponding colours. When the appropriate coloured button is pressed the decoder will immediately capture the pages corresponding to this topic, e.g. sport. The names on the bottom row instruction set would then change to, for example, football, cricket, tennis and swimming. Again, when the appropriate coloured button is pressed the pages associated with that particular sport are captured ready for display.

This technique has been developed as part of World System Teletext (WST) and is now called Fastext. The 'tree' structure of the Fastext page is illustrated in Figure 5.6. The editor can provide an escape from the branches by making one of the options (usually no. 4) lead back to the main magazine, or the viewer can use the page numbers.

An alternative system developed by the Institut für Rundfunktechnik (IRT) [15] in Germany, that does not require data packets is called TOPS. In this system the pages and magazines are arranged on a topics basis. A special control page, or 'table of pages' (TOPS) page, which defines all the other pages, is also transmitted. Each 8-bit character position in this TOPS page defines the category of each of the other transmitted pages. The

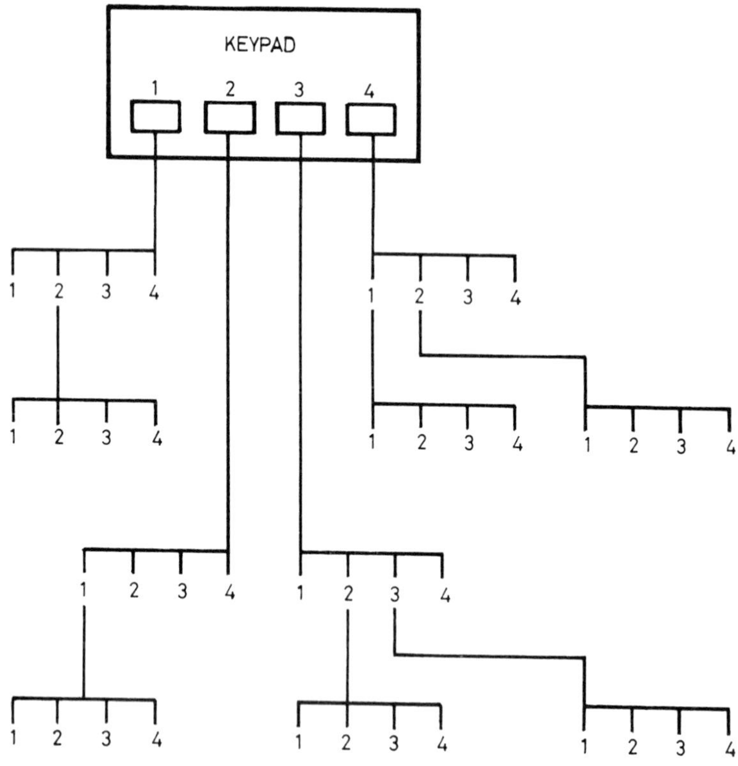

Figure 5.6 Fastext tree page structure

TOPS page is captured and held in the non-display decoder memory. Its function is to instruct the decoder processor as to which pages should be acquired ready for display in conjunction with the viewer's instructions received via the remote control system.

The magazines are organized using three basic page types: major topic pages, sub-topic pages and information pages. The handset has three corresponding buttons, together with one for reverting back to the previous page. When the receiver is first switched on, the decoder is programmed to display the first major topic page. The TOPS buttons on the remote control handset cause the decoder to display the next information page, or the next sub-topic page, or the next major topic page. To guide the viewer, the titles of the next sub-topic and major topic page are displayed on row 24. These titles are transmitted in additional information tables, which are contained in the nine information pages added to the transmission. The location of these information pages is contained in the page linking-table which forms part of the special TOPS control page that is used by the processor. The structure of

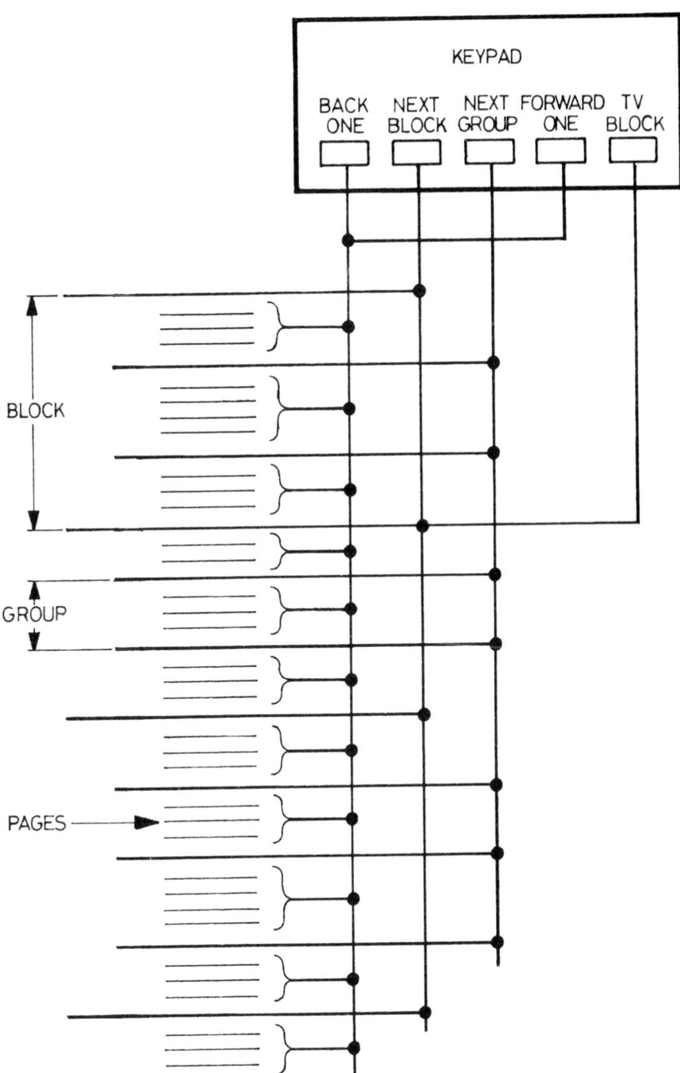

Figure 5.7 'TOP' data base structure

the 'TOP' database is illustrated in Figure 5.7. Page and magazine numbers are also used so that the teletext service can be used with receivers that are fitted only with conventional decoders.

The ideal receiver arrangement would be for the decoder to store the complete magazine of pages so that any page would then be instantly available, or, if storage is limited, for the processor to be programmed to automatically store the pages of the viewer's choice.

Teletext adaptors

Teletext adaptors are designed so that they may be added to a conventional television receiver to provide a teletext facility. The teletext adaptor normally has its own tuner, IF amplifier and demodulator driving a teletext decoder together with a remote control system. In fact it incorporates virtually all the circuits contained in a television receiver with the exception of the CRT and the associated time-bases. The RGB outputs from the teletext decoder are re-coded into a standard colour television signal (PAL, NTSC or SECAM). This signal is then modulated on to an RF carrier so that it can be fed into the aerial socket of a conventional television receiver. Such adaptors are not 'user-friendly' in that when the television channel is changed the decoder channel selector also has to be changed in sympathy, and a second remote controller is usually required for the adaptor. The requirement for two remote controllers can be eliminated by remodulating the video and sound signals onto the output carrier of the adaptor so that the television receiver then only needs to be tuned to the adaptor's output signal. Alternatively, if the receiver has a suitable video input it can be driven with RGB signals from the adaptor. In this way the adaptor provides the remote control of the television receiver but the video signal suffers in the re-modulation process when an RF feed is used to drive the receiver.

In general, the main limitation of teletext adaptors is the need to re-code the RGB signals from the teletext decoder into a standard colour coded television signal when no RGB input facility is available. Text signals are usually coloured and have fast edges. When such signals are converted into a coded colour television signal the bandwidth of both the colour and luminance components are limited, which causes a very significant loss of legibility in the final displayed text. Teletext adaptors have not proved popular due to the relatively high cost, and limited performance due to the re-modulation processes.

Control of video recorders by teletext

A standard video recorder contains a tuner with an AFC system, an IF amplifier and demodulator together with a remote control system. The output to the receiver is either an RGB signal or a remodulated RF composite video signal. The incorporation of a teletext decoder into the recorder therefore allows the unit to be used as a teletext adaptor with very little additional cost.

When the recorder is set up to record a programme at a specific time, it is assumed that the programme schedule will run according to the advertised times. In practice published schedules are often indicated only in 5-minute increments and variations may occur at the time of transmission, which means that the programme as recorded may be incomplete. This problem can be overcome if the broadcaster provides a control signal to activate the recorder in step with the broadcast programme. A signal for this purpose must be programme related in real-time that its presence or absence will activate the recorder, and it must also contain a code to relate it to the particular programme.

Such a control signal can be carried as teletext data within Packet 8/30, and very little additional circuitry is required in a video recorder that contains a teletext decoder to make use of this facility. Two data formats for such control signals are incorporated into the WST Specification [13]. The control code for either format is contained in Packet 8/30 and the data is maintained for the duration of the programme. In Format 1, the control code is a special 16-bit binary code number assigned to the programme by the broadcaster for each programme. This number is entered into the recorder by the user and the recorder is then left in the stand-by mode, i.e. the RF signal circuits and decoder are left switched on and tuned to the required channel. When the valid progamme identity number is received, the recorder switches into the record mode whilst the number continues to be received. This mode of operation requires the broadcaster to make available the special programme related code for each programme.

The Format 2 method has been specified by the EBU [16] to be compatible with the Video Programming System (VPS) which is already in use in a number of countries in Europe. The programme identification label is a 20-bit code which directly relates to the time of transmission as published. The user enters into the recorder this programme time in the usual way, and the recorder converts this automatically to the value which will match the identification label broadcast in Packet 8/30 for the duration of the programme. The programme can, in fact, be broadcast at any time as the recorder will only automatically record the programme when the appropriate Packet 8/30 is received. It is envisaged that where fully integrated or linked television receivers and recorders are available the special programme code will be generated from the teletext television programme guide pages. The user will have a cursor control facility on the keypad and an 'enter' button. When the cursor is moved to the required programme time, pressing the 'enter' button will cause the corresponding programme code to be generated, which together with information concerning the television channel etc. is entered automatically into the recorder. The recorder control logic can assist the user by drawing attention to potential time conflicts between recordings and also the duration of the tape required. At the broadcasters' premises, the necessary programme information for generating the Packet 8/30 code is automatically provided to the teletext system from the programme scheduling computer.

In-vision systems

A teletext system can provide a source of input pages for an information service on either a broadcast or a cable television network. This arrangement is generally referred to as an 'in-vision' facility. The teletext page is decoded to RGB signals and re-coded into a PAL (or SECAM, or NTSC) colour signal so that it can be received by normal television receivers. The different pages are transmitted cyclically, each one being transmitted for a display time of about 10 to 15 seconds depending on the content. A series of pages can also be transmitted very rapidly so as to provide an animated display sequence in the programme. The pages can be specially created or the service can use any normally available teletext pages. The teletext decoder is equipped with a processor and page memory of adequate capacity for the service, typically 30 to 100 pages. The processor is programmed to select the required pages from the teletext input and hold them in memory ready for transmission. Each page is then 'enabled' for transmission for the required display time. Simple in-vision decoders usually use the same display time for each page, but more complex decoders allow the display time for each page to be varied and this is set when the page sequence is chosen for the service.

The wideband RGB signals are matrixed by the PAL coder into a wideband luminance signal and two narrowband colour-difference signals. Decoded teletext signals have fast edges, as in a receiver they are fed directly from the decoder to the video output stages. If the RGB output signals from a teletext decoder were coded into a composite video signal, transmitted over a broadcast or cable system, received, demodulated and decoded into RGB signals they would suffer considerable degradation. This degradation of the signal is acceptable for normal television pictures but it makes a text display very difficult to read. The vertical elements of a character consists of very narrow pulses which are reduced in amplitude and have very poor coloration; whereas the horizontal elements, which are represented by much wider pulses, are reproduced at full amplitude with good coloration. Furthermore, the reproduced saturation of the page background colour may not be optimum for the text colours.

The legibility of the text page can be greatly improved by text enhancement before the PAL coder. The width of the narrow vertical pulses, which form the vertical elements of the characters, are stretched to reduce the bandwidth, and the saturation levels of the background colours are reduced. The character shapes available from more recently produced decoder circuits have wider vertical elements, therefore this effect is less pronounced. The background colours can be reduced in amplitude by the processor software. When adjustments are being made to an in-vision decoder to improve the legibility of the final pages, it is important that the pages are viewed on a television receiver using an RF input, so that compensation can take into account all the potential distortions that occur in the complete system.

6 Networking and regional services

The teletext data signal parameters were chosen and specified so that the data could be inserted in the field blanking interval of the normal television signal without disturbing the normal vision and sound signals. The video signal can therefore be considered as a common carrier for the data signal. The data signal, consisting of high speed pulses, is sensitive to amplitude, group delay and non-linear distortion. Such distortion causes overshoots and a deterioration in the separation between the '0' and '1' levels of the signal.

When a television signal is distributed by a wide spread distribution network, it passes through various links and switching centres. Although the quality of the colour television signal is maintained to broadcast standards, the data signal can suffer some degradation. To provide the maximum teletext service area for a given transmitter and also to ensure that the data signal does not disturb the reception of the transmitted television sound or vision signals, it is essential that the data signal be radiated without degrading the pulse shape, and the amplitude of the data pulses must not exceed the correct value. If the amplitude of the data pulses is low, then teletext reception will be impaired and is more likely to be further impaired by poor received signal quality and noise. On the other hand, if the amplitude is high teletext reception may be improved, but receivers are liable to exhibit 'sound buzz' caused by the excessive amplitude of high frequency components of the data signal interfering with the inter-carrier sound signal.

The digital data signal can be processed independently of the video signal. This enables the data to be regenerated or linked to other television networks as required for the teletext service, without disturbing the normal television network.

Data bridging

Television networks often have many special arrangements for distributing video signals which can vary during the day depending on where programmes have been originated and which transmitters are required to radiate them. This is particularly the case for television services which have different transmitters for different programmes during the day. The teletext service, however, is likely to be produced centrally, therefore the teletext signal needs to be distributed throughout the television network independently of the studio signal sources.

The teletext data signal can be transferred between different parts of a network using a 'data bridge'. The data bridge has two video input circuits, each of which provides sync, blanking and data clock pulses. The teletext data signal from the A channel is separated from the video signal, error-corrected, and clocked into a buffer memory as shown in Figure 6.1. The memory capacity must be adequate for buffering all the teletext data that is contained in one FBI period of up to 16 lines, that is 16 × 360 bits. If field integrity is required to be maintained then double this capacity is required as the data must be stored for two field periods. The teletext data need not be inserted on consecutive FBI lines as the insertion test signals (VITS) are inserted onto agreed lines and the teletext data can be inserted on lines either side of such signals. It is therefore necessary to maintain the teletext data line integrity, that is the data on line 21, say, is always bridged onto line 21, irrespective of teletext signals on other data lines. Maintaining data line integrity ensures that page headers (row 0) for new pages, subtitle data on dedicated lines or data services using teletext format do not become line shifted.

The memory output circuit is controlled by signals generated from the B channel. The output is serialized and bandshaped to the required specification. This is achieved by passing the fast digital bit stream, which has identical rise and fall times, through a special band shaping filter. Teletext data is then inserted onto the B video channel, the video lines being erased prior to insertion to remove any noise and thus ensure the highest possible quality of the teletext signal on the second network. The two video signals can be asynchronous but the number of data lines on each channel must be the same. The line numbers, however, need not be the same. A teletext signal can therefore be passed through very complex video television networks or can bypass studio centres as required, without any disturbance to the normal television operations. Furthermore, the data is completely regenerated by the bridge at the point of transfer to full specification.

Whilst at first sight it might appear that the input requirements for a data bridge are less stringent than those for a receiver, they are in fact rather more critical. The data bridge provides the input

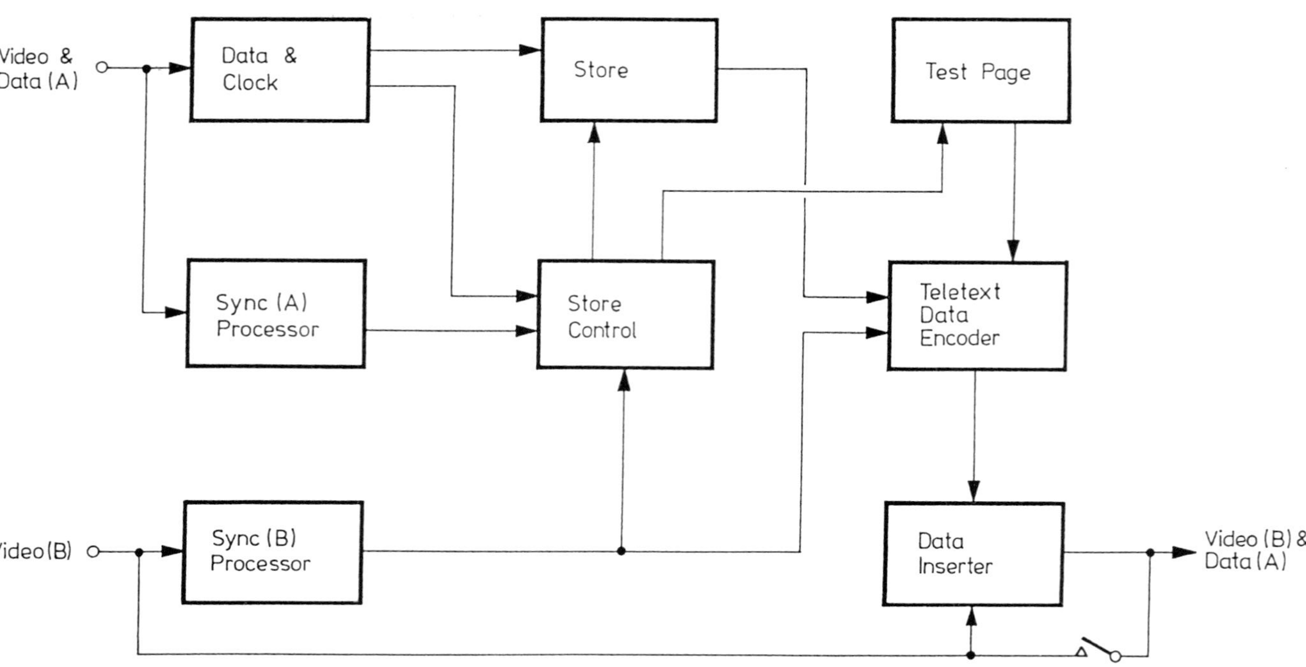

Figure 6.1 Functional diagram of a data bridge

teletext signal to the television network which will then be transmitted to a very large number of receivers. It is therefore imperative that the data signal meets full broadcast specification at the bridge output in spite of signal variations, sync genlocking and other phenomena that occur in television networks. It follows that the data inserter must also meet full broadcast specifications and under no circumstances must the bridge insert data during the actual video signal period.

The data bridge incorporates a test page generator which is initiated whenever no data signal is received from the input network. The page can be programmed to contain an apology caption, which can include the location of the bridge in the network, and it can also maintain test signals at all times for the use of technicians installing or maintaining teletext equipped television receivers.

The data bridge can also shift the teletext signal to different television lines. Thus various teletext sources may be combined onto one video signal, which then acts as a carrier for distribution purposes.

The normal teletext signal consists of 7 bit characters plus an odd-parity bit. Although each character is then represented by an 8-bit code, the code will always be of some combination of 0 and 1. These data transitions are often used in decoding circuits to help maintain the phase of the regenerated clock. When software code or computer data is being transmitted characters which consist entirely of 0s or 1s may occur and transients are not then available to help maintain the phase of the regenerated clock. The transmission of such 8-bit code therefore requires special tests to check that equipment in the television network will not produce errors. This aspect is of particular importance for data transmission to 'closed user groups' as the data is not repeated and an error in transmission could invalidate the message (Chapter 9, page 106).

Data regeneration

Some distortion inevitably occurs in the transmission of television signals in widespread networks. The specifications for video links, and associated measurement techniques, are designed to ensure that the television signal meets broadcast standards. A somewhat different weighting of distortion criteria is appropriate for fast data signals which are particularly sensitive to group delay distortion and to the non-linear distortion produced by some re-broadcast links (RBL).

Provided the teletext data can be decoded correctly, its characteristics can be fully restored by regeneration. At the present time, regeneration of the data signal must be carried out at video. Regenerators can therefore be used only where a base band signal is available, such as at the input to a transmitter.

The regenerator strips the data from the video signal, completely reprocesses and retimes it, carries out band shaping and then reinserts the data onto the video signal one field or one frame period later. The video line is 'erased' prior to insertion. The video inserter section of the regenerator must meet the full colour television broadcast specification as it is in the main programme path, and it must also be equipped with a physical bypass relay which operates in the event of a malfunction. The output circuits of a regenerator are therefore similar to the output circuits of a data bridge.

In addition to their use at transmitting sites, regenerators are also necessary at the outputs of video tape machines to restore and line shift the data when teletext subtitles are recorded together with the programme material.

Data regenerators cannot be used at transposer stations because the vision signal is converted only to an intermediate frequency, not to base band. Current transposer designs are broad band to accommodate combined sound and vision signals and, since they have no sound notches, they do not introduce significant group delay errors at high video frequencies. The need for regeneration equipment in transmission chains incorporating transposers is therefore much less as the data signal degradation is smaller.

Standards conversion

A teletext signal is independent of the video signal which acts as a common carrier. In a 625-line/50 Hz teletext system, 40 characters are carried in one data line period. There are, in addition, the clock run-in, framing code and row numbers, so that the total number of 8-bit characters per line is 45. The data rate for one data line per field in a 625-line system is therefore $45 \times 8 \times 50$, i.e. 18 000 bits per second. The bandwidth of 525-line/60 Hz television systems is narrower and only 32 characters can be transmitted in one television line period, which has approximately the same time duration as that of 625-line television systems. The corresponding data rate is therefore $37 \times 8 \times 60$, i.e. 17 760 bits per second. The data rates for 525- and 625-line television systems are therefore substantially the same, the 525-line system having a slightly lower data rate.

Teletext transcoding from one television standard to another, in either direction, therefore requires a unit rather like a data bridge but with the input and output circuits each operating on a different standard. The data rates for the 525- and 625-line systems are similar and the displayed page format is the same (40 characters per row and 24 rows per page). However, transmission format is different, as in the 525-line system only 32 characters are transmitted in one data line, the remaining eight characters of each of the preceding four rows are transmitted together as a separate data line. So four rows require five data lines to transmit and each page of 24 rows therefore requires 30 data lines. The buffer memory needs to have a storage capacity of at least one page. In conversion from 525-line to 625-line teletext, the input data rate is 240 bits/second, or 30 characters/second, lower than the output data rate. It might therefore be expected that the 625-line teletext output signal would occasionaly have blank lines, when there was no data available for immediate output. However, the transmission of blank lines can give rise to misleading reception and measurements in a broadcast system and the normal practice to avoid this is to duplicate a line of data.

Conversion from 625 lines to 525 lines requires the buffer memory to be capable of storing the complete transmitted magazine of teletext pages, as the output data rate is lower than the input. The 625-line input signal is decoded and the pages are written into the buffer store on a continuous basis. The output of the store is under the control of the 525-line signal which continuously reads from the memory and outputs the data at the 525-line rate.

There are a number of advantages in having a memory capacity able to contain the complete teletext magazine. It enables a different number of television lines to be used at the output from that of the input and, in the event of the input signal being switched off, the output is maintained. The input data signal updates the contents of the memory and the contents are then continuously available for the output. The converter can therefore increase the number of transmission lines used by the output system and the access time at the output of the converter can be very much faster than that of the input. The real time clock would normally be generated by the processor used in the converter so that the correct local time may be incorporated into the output page headers together with other station or service identification.

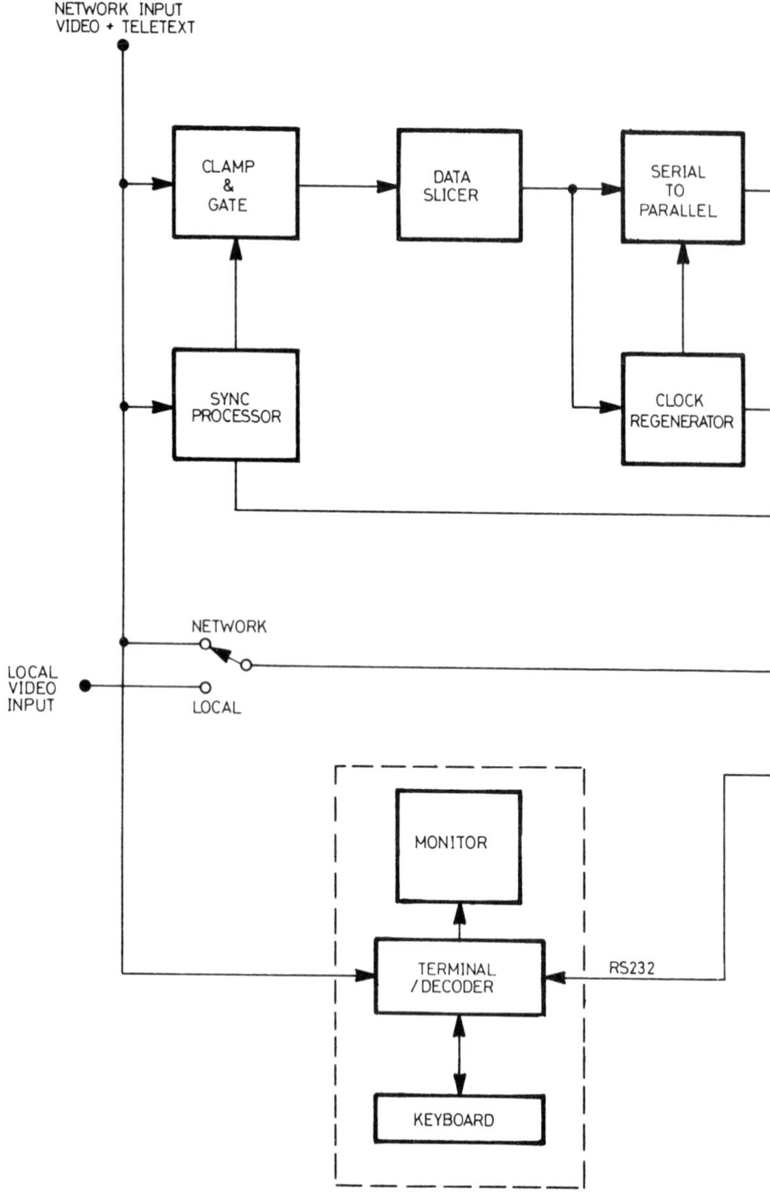

Figure 6.2 Page exchange system for regional teletext services

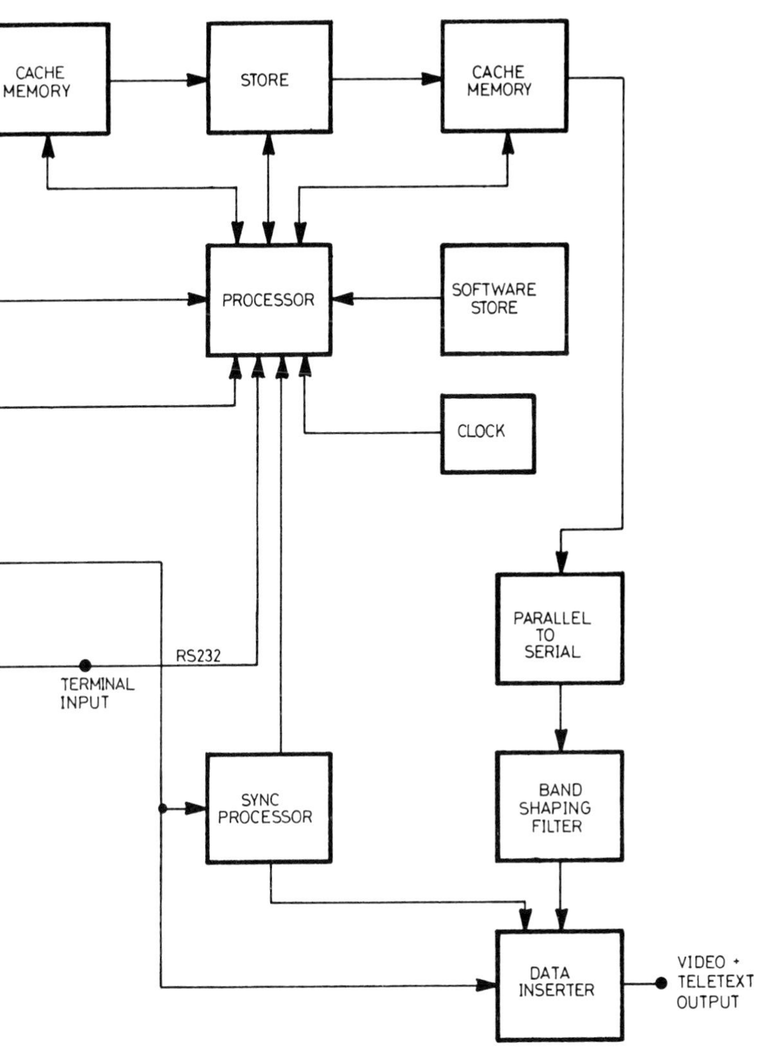
CACHE MEMORY
STORE
CACHE MEMORY
PROCESSOR
SOFTWARE STORE
CLOCK
PARALLEL TO SERIAL
RS232
TERMINAL INPUT
SYNC PROCESSOR
BAND SHAPING FILTER
DATA INSERTER
VIDEO + TELETEXT OUTPUT

Regional services

Regional teletext services can be provided in a number of ways depending on whether the pages are to be edited locally or at a central point.

A separate teletext system operating in parallel with the main system can provide a completely separate local service provided there is an unused FBI line available and a magazine number not used by the main service. Both teletext systems must operate in the parallel mode to prevent decoders being confused by the data signals from the two teletext systems. A control bit (C11) in the page header is set to '1' for serial magazine operation so that the decoder display clock is updated by the pages in any of the magazines. If this bit is set to '0' for parallel operation but the magazines (on the main system) are being transmitted in series, then the decoder can respond only to the data of the selected magazine. The display of clock time will then only be updated during the transmission of this magazine. A regular update of displayed clock time can be achieved by interspacing pages (or page headers), from the other magazines in the series.

The arrangement used for page exchange is shown in Figure 6.2. The signal from the video input is decoded, error checked and fed to the data store. This store provides the buffer memory for the main teletext pages and also the store for locally created pages. The system processor controls the read function under the control of the resident software. A data entry port is provided so that the editing terminal can communicate with the system. The editor creates the pages that are to replace others in the magazines. When these pages are entered into the memory, they must be given page and magazine numbers corresponding to those that they are to replace. As the input pages from the main programme are decoded, the page numbers are checked by the processor, which is then able to insert the local pages into the teletext output on an exchange basis.

As the pages of the main teletext service are not stored – only the current page being held in the buffer memory – the replacement pages must not require any more rows of data than those they replace. The local editor must therefore be able to check all the aspects of a page that is being exchanged. This is particularly important when additional rows or packets are used, as required by certain languages for example. It is also essential that the same number of television data lines are used at the output. These problems do not arise if the complete teletext magazines of the main programme are stored. Furthermore, the output can be maintained if the input signal is temporarily disconnected, and the number of data lines can differ from the input. Page exchange facilities allow a local or regional service to become operational quickly and the local facilities can be expanded later as required.

The editing terminal needs to contain a teletext decoder and to have a direct feed from the input video signal. The local editor can then capture any page from the input for examination at any time, independently of the local system. This facility is essential when the system does not store the complete input magazines of pages.

The video signal acts only as a carrier for the teletext data and the video programme content is not relevant except when there are teletext subtitles. In such cases, if the local video programme is different, the subtitles must be deleted from the teletext magazine, and possibly stored with the network video programe material for later use (Chapter 7).

7 The subtitling of television programmes

Teletext allows the transmission of subtitles (or captions) that need to be displayed only as and when required. Subtitles transmitted in this way are referred to as 'closed captions' whereas subtitles inserted as a video signal into the programme video are called 'open captions'. Teletext subtitles are of obvious benefit to viewers with hearing impairments. They may also be used to make available simultaneously subtitles in several different languages, by transmitting them on different pages. Preparation of subtitles for transmission with a prerecorded programme is a very time consuming process but the use of microprocessors and software incorporated into specially designed preparation systems has reduced the time required from about 30 hours to some 10–15 hours per programme hour.

Subtitling of prerecorded programmes

Subtitles for prerecorded programmes are prepared in advance of transmission and stored on floppy discs. The subtitle editor would normally have the programme to be subtitled available as a tape recording together with a programme script. This tape 'dub' of the programme master tape will also have the studio time-code reference which is used for programme production purposes. The time-code can therefore be used to key the subtitle 'on' and 'off' times. Subtitling takes place separately from programme production and also separately from the technical and editorial procedures concerned with the provision of a teletext service. Subtitle preparation can therefore be undertaken with purpose-designed work stations in a separate location.

The functional diagram of a typical station is shown in Figure 7.1. The programme to be subtitled would normally be viewed from a tape dub with time-code copied from the programme master tape. A subtitle script would normally be prepared by manually editing the programme script to a reading rate of about

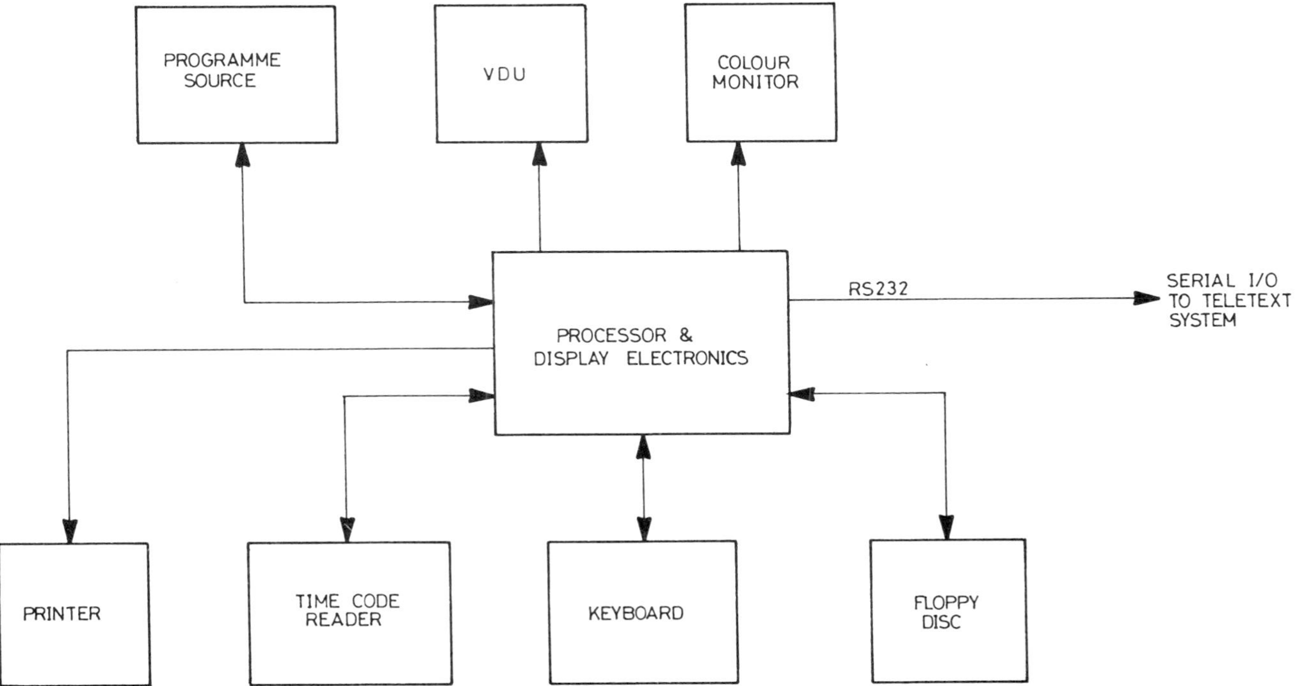

Figure 7.1 Subtitle recording system for regional teletext service

100 to 120 words per minute. In preparing the subtitle script, key words of utterance are preserved where possible; at the same time the sentences are made more straightforward and therefore easier to read quickly. Supplementary information relating to background noises or effects not obvious to a viewer with a hearing impairment is also provided. The editor aims to convey sufficient information for the viewer to appreciate and enjoy the programme, without imposing an unreasonable reading load. Review of the programme tape with the script enables notes to be made regarding subtitle placement. It is helpful for the deaf viewer if subtitles are laterally positioned on the screen to match the position of the speaker. Further assistance is given in identifying a speaker by using differently coloured subtitles for different characters where possible. Notes can also be made on the script to indicate the time available for each segment, which will guide the later editing procedures. The editor prepares the subtitles using the keyboard and VDU display. Each subtitle is individually numbered and stored on the floppy disc. When all the subtitles have been entered for the complete programme, it is then necessary to enter the time-code which will specify when the subtitle is to be displayed, together with the time-code for cancelling the display. Thus two time-codes must be entered per subtitle. In an alternative method the processor calculates the necessary length of time for display of the subtitle, for a given reading rate, and automatically adds the appropriate 'erase' time-code. A print-out of the subtitle and associated time-codes can be produced. A typical print-out is shown in Figure 7.2. The programme would be pre-viewed on a colour monitor which would also have a display of the time-code generated by the time-code reader.

The method of subtitle preparation described allows for maximum flexibility. For example, a number of television programmes can be worked on concurrently, the subtitles being viewed independently of the teletext system, and in the order in which they will eventually be transmitted. If an amendment is needed to either the text or to the associated time-code, flexible editing facilites are available for corrections to be made. A floppy disc is a very convenient medium for storing a complete programme of subtitles with associated time-codes. However, floppy discs do impose restrictions on the editing arrangement since if any major editorial changes are made to the television programme, the complete disc has to be reprocessed. Furthermore, instant access to any part of the programme is not available to the editor.

Time-code	Subtitle Text	Colour	Display Feature
00 11 05 02 3*	Anne, sorry to call you at home.	C	
00 11 06 24 4*	Can you arrange to meet me at lunchtime?	C	Up 1
00 11 09 12 2*	Yes, of course.	Y	Add
00 11 11 19 2*	DOOR BELL >	W/R	Effect
00 11 14 00 3*	Hold on, someone is at the door.	Y	

Time-code — hours, minutes, seconds, frames (hh, mm, ss, ff).
* Display time.

Colour — C = cyan, Y = yellow, R = red, W = white.
W/R = white text on red background etc.

Display feature — Up 1 = subtitle moves up 1 line.
Add = added to existing text (add-on).
Effect = off-screen effect; arrow indicates position.

Figure 7.2 Typical subtitle print-out

A RAM disc is used in the more recent designs of subtitle preparation systems, so that all the subtitles for the complete programme are instantly available. When a programme has been completely subtitled, the contents of the RAM disc are transferred to a floppy disc for archival or operational purposes.

Subtitling of live programmes

As previously mentoned, preparation of subtitles in advance for a prerecorded programme usually now takes some 10–15 times the programme duration. During an unscripted 'live' broadcast, the subtitles must be composed, entered and formatted as the programme proceeds and it is almost impossible to achieve the same presentation standards as can be achieved when subtitles are prepared in advance. The problem becomes one of trying to present the viewer with sufficent information quickly enough for it to be in context, and to present it in a form which is easy to comprehend. These tasks can cause serious problems for the editor, particularly during a rapid live presentation of unscripted material. However, in practice, few programmes are live and completely unscripted with no information available beforehand about what will be said or who is participating. Portions of live programmes are often prescripted, particularly coverages of major public events and parts of news bulletins. The subtitle preparation system normally contains a file of advanced information relating to the event to assist the editor. Such files are normally called 'shortforms' and are accessed by a simple code. A shortform may contain two- or three-word names or phrases that are expected to occur frequently in a programme. For example, in a sports programme the individual players' names would be stored in full, but accessed by a number. Before transmission a list of such names and phrases would be produced as background knowledge and information likely to be required during the programme. Appropriate abbreviations are chosen by the editor and stored with the full names as a dictionary in the subtitle preparation system. When a shortform is later required as part of the text of a subtitle, the processor automatically substitutes the longform from the stored directory in response to the simple code entered by the editor.

To further assist in subtitling it is common practice that two or three keyboards are able to access the processor simultaneously so that two or three editors can be involved at once. This technique is particularly valuable on sports type programmes as one editor can follow the play while the second editor can provide backup information or undertake subtitle preparation, as appropriate. An additional display monitor may be incorporated into the system so that shortform or other prepared notes can be made available to the editors without complicating the VDU display.

The use of a machine shorthand keyboard in conjunction with a specially trained operator can provide a faster input with certain languages. Phonetic codes representing speech are entered and these are decoded by the keyboard computer to produce a conversational transcription. Opinion is varied as to the value of such systems as they can produce ambiguous words or sentences under certain conditions; but the advantage of machine shorthand input is that a rate of about 100 words per minute can be sustained relatively easily.

Subtitle transmission

A subtitle page cannot simply be inserted according to the page number in the teletext magazine. If this were done, the page would be transmitted anywhere between 0 and 25 seconds after insertion (assuming a 100-page magazine using two television lines). Even inserting the subtitle at the end of the current page is inadequate, since the transmission delay would then vary between 0 and 0.25 seconds. It is therefore necessary to start transmitting the data for the subtitle on the first available data line after the cueing point. Special techniques must be used in the teletext system to give priority to the subtitle input and thus to allow the page to be inserted into the magazine in this manner.

Transmission of subtitles as part of a teletext magazine also places major limitations on television networks, particularly where the programme material on regional sections differs from that on the main network. Furthermore, regional stations often transmit network films (or other material) at times different from the main network. Consequentially, a programme on the main network will have subtitles at the wrong times for regional transmission. It is therefore necessary to intercept the subtitle pages at the regional station and to record them for transmission at the later time, with appropriate programme material. The equipment necessary for this purpose is shown in Figure 7.3. It is an extension of equipment required at a regional station for page exchange purposes (Chapter 6, page 70).

Referring to Figure 7.3, on record, the VCR is fed with the programme video signal and the teletext subtitle page only. The processor is programmed to output only the subtitle page, which is also line-shifted to occupy the last FBI line before the video signal. This is to prevent the VCR erasing the teletext data from the FBI. The data is inserted at full amplitude to help overcome limitations arising from shortcomings in the VCR's frequency response.

On replay the teletext subtitle signal is decoded, then applied as an RS232 feed direct to the teletext system subtitle input port. The system software is programmed to give priority to the subtitle input port so that each subtitle is inserted into the magazine on the next available FBI line.

An alternative arrangement that avoids many of these

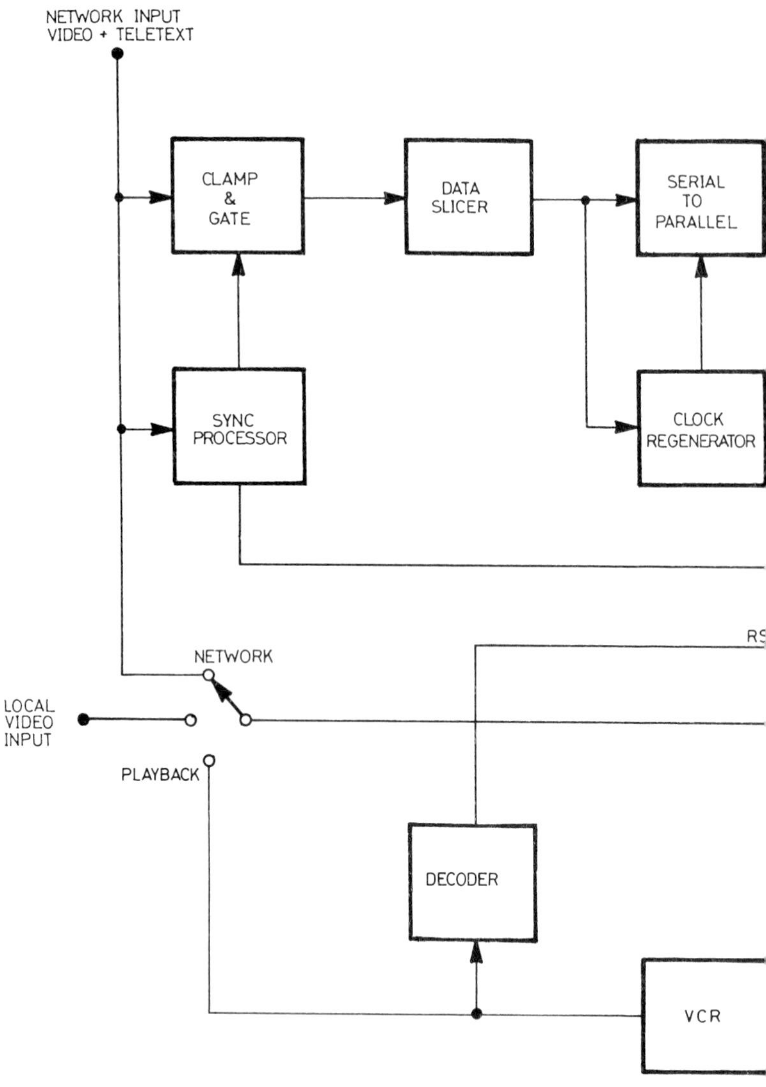

Figure 7.3 Subtitle recording system for regional teletext service

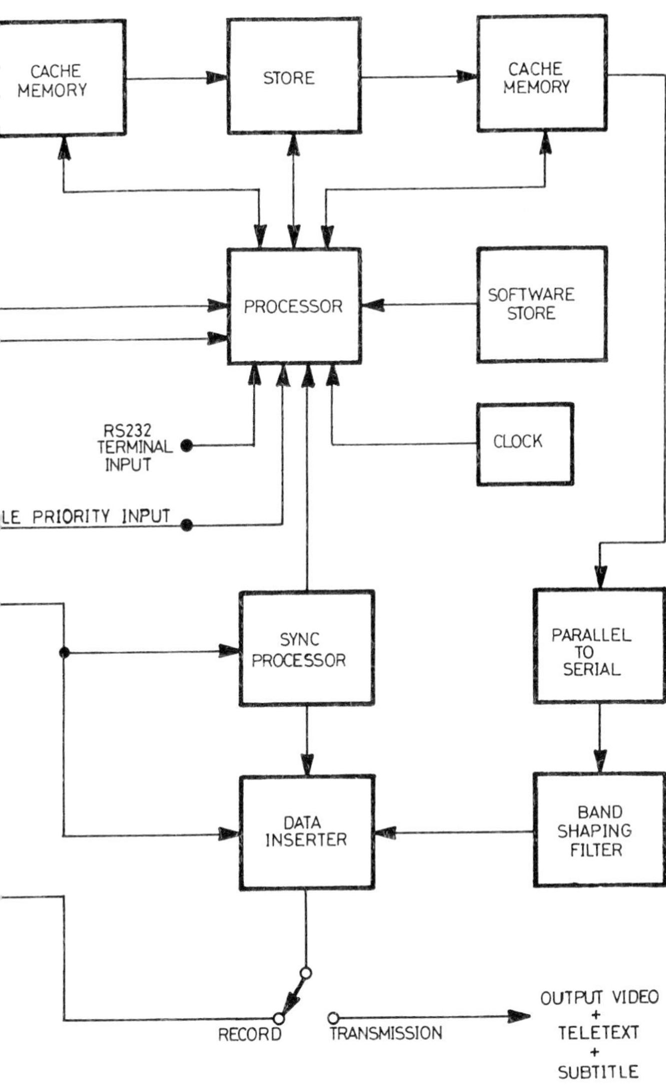
CACHE MEMORY
STORE
CACHE MEMORY
PROCESSOR
SOFTWARE STORE
RS232 TERMINAL INPUT
CLOCK
LE PRIORITY INPUT
SYNC PROCESSOR
PARALLEL TO SERIAL
DATA INSERTER
BAND SHAPING FILTER
RECORD
TRANSMISSION
OUTPUT VIDEO + TELETEXT + SUBTITLE

complications is to devote a separate FBI line to the teletext subtitles. The subtitles can then be transmitted completely independently of the main teletext service. A further advantage of a separate FBI line is that regional stations, or other users, may use the subtitle information in the FBI with the video signal, independently of the teletext service. This is particularly important when the regional station is tape recording the video signal for transmission at a different time.

The use of a separate FBI line allows the programme to be simply recorded with the subtitle. When the programme is transmitted the subtitle is generated directly from the tape recorder along with the video signal. The only additional data processing required is the regeneration of the teletext data to compensate for the shortcomings of the recorder, and possibly the shifting of the data onto the correct FBI line for transmission. This line-shifting facility is a normal feature of a teletext data regenerator.

The use of a separate FBI line also simplifies teletext networking since the subtitle FBI line can be treated as programme-related information and is never separated from the video signal. It also overcomes the problem of floppy discs which contain the subtitle information becoming separated from programme tapes, or even being associated with the wrong programme tape.

8 Satellite data broadcasting

Direct broadcasting by satellite (DBS) systems use frequency modulation (FM), rather than amplitude modulation (AM) as used in most terrestrial systems, because of the limited power of the satellite transmitter. Some DBS services use conventional PAL, or NTSC, for coding for the colour television signal but services introduced recently employ the MAC (multiplexed analogue components) system [17].

With conventionally coded signals the teletext data can be carried in the FBI, as for terrestrial transmissions. The received decoding margin is somewhat lower in the case of DBS due to the non-linearity of the FM demodulator in the receiver, but the decoding margin of the demodulated signal is adequate for normal teletext decoders. Furthermore, the satellite aerial system is highly directive and therefore the signal is not normally contaminated by reflections, which are most often the main cause of distortion of digital data signals.

The MAC transmission system uses time multiplexing for the luminance and colour-difference signals rather than frequency multiplexing as used for conventionally coded colour signals. Digital techniques are used to time compress the luminance signal in time by a factor of 3/2, and the lower bandwidth colour-difference signals by a factor of 3. The compressed luminance signal and one colour-difference signal are transmitted in one line period. The other colour-difference signal is on alternate lines. At the receiver the video signals are expanded and the RGB drive signals reconstituted from the luminance and colour difference signals in the conventional manner. If a digital teletext signal were included in the FBI periods of the original video signal it would be time compressed with the luminance signal. At the receiver, the data signal is expanded but unfortunately the compression and expansion of the data signal results in a very significant reduction in the decoding margin, typically some 40%, which renders the data signal unusable by present teletext decoders. The teletext data signal in the FBI period could bypass the compression circuit in the coder, since there are no colour-difference signals present, and use the full line period. A delay circuit would then be necessary in the receiver to compensate for the delay in the video processing circuits.

Although digital compression techniques are used for the video signals, all the other signals such as stereo sound, line and field synchronizing, blanking and the data are all sent as data packets. The format of the transmitted MAC signal is shown in Figure 8.1. The specifications for the MAC transmission standards detail the location and format of the different data packets together with the priority required for different purposes [18]. A number of variants

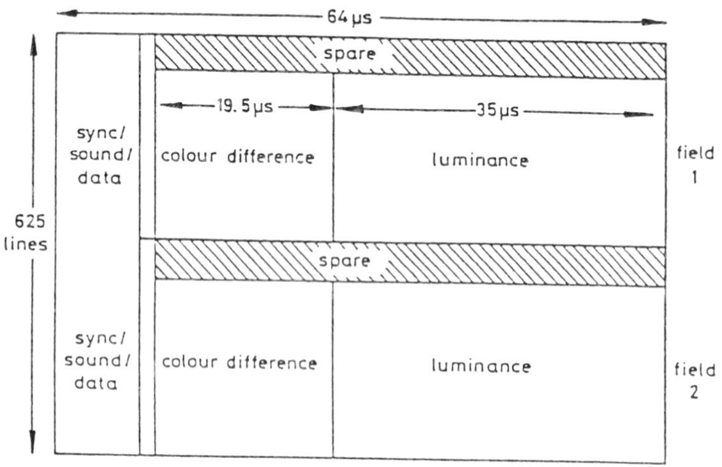

Figure 8.1 Transmitted frame for MAC signal

of the MAC standard have been specified which have differing data capacities. Nevertheless all of them provide adequate space for a teletext service. The teletext data is conveyed in MAC packets which contain the address header followed by the useful data. Two possible levels of error protection are provided, the lower level corresponding to that used in normal terrestrial systems. The higher level of protection, with its larger overheads, halves the useful data that can be carried by each packet.

It is likely that many satellite receivers will be adaptors for use in conjunction with existing television receivers. If the television signal is transcoded to PAL format the teletext data can also be transcoded to a conventional teletext signal format in the fbi for use with a standard television receiver. Receivers that employ an integral MAC decoder would not require the data transcoding facility as teletext decoding would be part of the MAC decoder function.

When the MAC transmission standard was being established a number of different data coding techniques were considered. A proposal, known as 'C-MAC', employed a form of coding with a 2–4 phase shift key arrangement (symmetrical PSK). This method is technically superior to the other proposed formats, providing very good error performance in the presence of noise and making full use of the satellite channel bandwidth. Nevertheless, after further consideration, a duo-binary coding system [19] was found to be a highly acceptable compromise as it provided an economy of baseband bandwidth and commonality with other proposals made to the EBU. A duo-binary (three-level) coding technique permits a data rate of 20.25 Mbits/second, which corresponds to the video sampling frequency in the 8.5 MHz baseband channel. This method is used in the 'D-MAC' system. The 'D2-MAC' system provides half this data rate. 'B-MAC', which is used for the network distribution of television and data signals by satellite, uses quartenary (four-level) coding and provides a data rate of 14 Mbits/second.

The essential feature of a duo-binary coding system is that it is a three-level data format, as illustrated in Figure 8.2. Within a given bandwidth, this format provides a data rate which is twice that of a two-level system, but the noise threshold is halved. A binary-coded signal can be translated into duo-binary code by passing it through a relatively simple delay and differential coding circuit, as shown in Figure 8.2. The input data is inverted and applied to a differential coding circuit. The output (waveform 3) is 1 only when there is a difference between the two inputs to the exclusive OR gate and one input is a 1. This output is delayed by 1-bit period and added to it to produce the three-level duo-binary data. The 1 in the input data always corresponds to a +1 or −1 in the output, the input 0 is always at the output 0 level. (Had the input data not been inverted then the input 0 would correspond to the output 1 and the input 1 to the output 0 level.) If an odd number of 0 has occurred between the current and last 1 in the output, then the output has a different polarity.

A typical decoding arrangement is shown in Figure 8.3. The output of two slicing circuits is applied to an OR gate so that the output data is the sum of the data contained in the two levels of the input signal.

The data capacity of the sound and data packet multiplex of the D-MAC or D2-MAC packet systems can be divided in a flexible

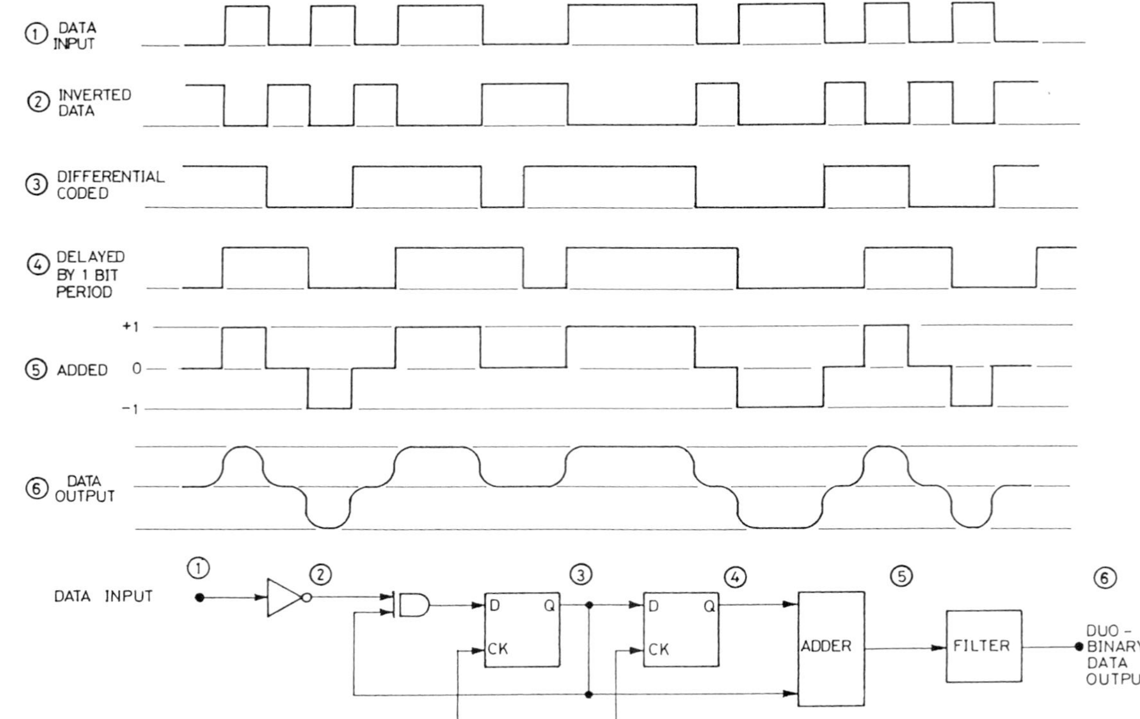

Figure 8.2 Duo-binary coding of data

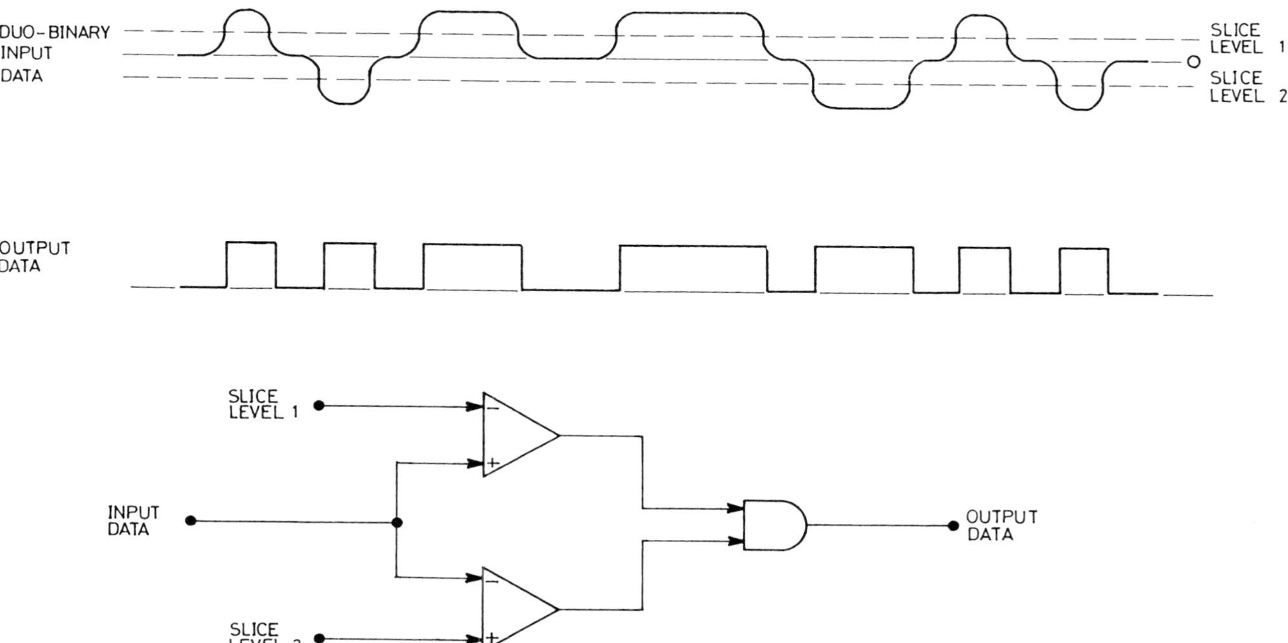

Figure 8.3 Decoding duo-binary data

way between sound channels, the service identification channel and other broadcast data services such as teletext, subtitling and telesoftware, together with a transparent data channel. If a vision signal is not required then the data capacity of a MAC packet channel can be increased by replacing the area of the frame normally occupied by the vision signal (Figure 8.1) with data bursts composed of data packets, in a similar manner to the existing data burst.

Teletext services using the 'D-MAC' system are carried in the data packets. The format of the teletext data is similar to that employed in WST in which 45-byte teletext rows or packets are used. (These should not be confused with MAC sound/data packets.) The first three bytes of the 45-byte block comprise a clock run-in sequence followed by a framing code, which provides synchronization of the teletext data when carried in the FBI of terrestrial television signals. These first three bytes are not required for recovery of the data from MAC packets and are therefore not included in the MAC teletext data block. The remaining 42 bytes consist of a 2-byte magazine and teletext data packet address group (MPAG) followed by 40 bytes of teletext characters. A control byte (CB) is added and then a 2-byte cyclic redundancy check (CRC) which covers the 40 bytes of the teletext characters within the block. The data in both MPAG and CB are Hamming coded for error protection and the structure of the teletext block is illustrated in Figure 8.4.

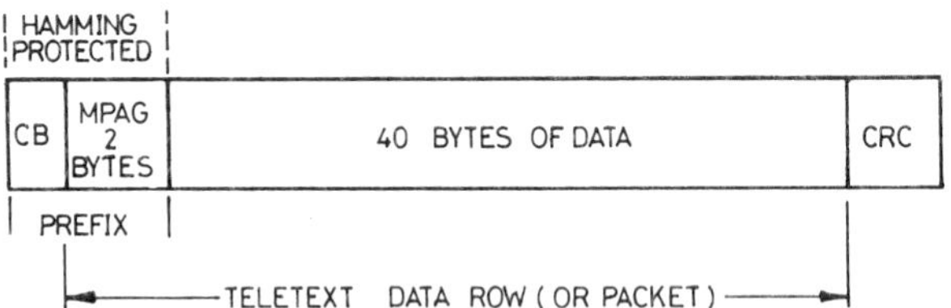

Figure 8.4 Teletext data block

Two types of MAC packet are specified for conveying teletext information in the sound/data multiplex, each having different levels of error protection. The first level of protection is intended for use with conventional teletext data services where any error correction can be achieved from the repeated acquisition of the data. The repetitive information in the control byte, together with the CRC code on each 40-byte data field, allows majority logic or bit variation techniques to be used for correction of errors in the block. These arrangements are particularly useful when 8-bit data is being conveyed, for which there is no parity protection on individual bytes. The first level protection of the MAC packet comprises two teletext data blocks carried in the 90-byte useful data field of the MAC packet, as shown in Figure 8.5(a).

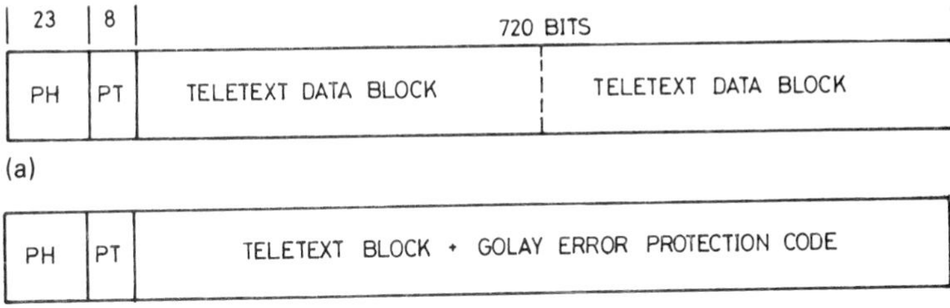

Figure 8.5 Teletext data blocks in MAC-packet. (a) First level protection, two teletext blocks per MAC packet; (b) second level protection, one teletext block per MAC packet

For the second level of protection only one teletext block is carried in each MAC packet. The entire 45-byte teletext data block is coded using the (24,12) extended Golay code. This provides a high level of forward error protection. The 90-byte useful data capacity of the MAC packet therefore consists of 30 Golay words, each containing 12 bits of the teletext data block, as outlined in Figure 8.5(b). The forward error correction of the second level packets affords a significant increase in ruggedness of the data in the presence of errors, so that repetition of data for error correction purposes may be reduced or eliminated.

As for other packets within the MAC sound data multiplex, the packets carrying teletext data commence with a 23-bit packet header (PH) followed by a packet type (PT) byte. The packet header comprises a 10-bit packet address, a 2-bit continuity index and an 11-bit protection suffix provided by a (23,12) Golay code. The continuity index is incremented on successive packets with the same address. For 'controlled access' to the teletext packets the PT byte indicates the type of scrambling that has been applied to packet data. There are three possible values for the PT byte in this application, corresponding to free access scrambled, controlled access scrambled and unscrambled data.

The existence of teletext services in the MAC sound data multiplex is signalled in the service identification (SI) channel. The packet address and sub-frame location for the teletext services are given by an entry in the list of indices carried in data group 0 of the SI channel. This locating information may be supplemented in a subsidary data group, by the name and language of the service and details of any related service features (for example, conditional access).

A teletext service implemented in conjunction with a MAC packet transmission system would utilize conventional teletext data management facilities, as described in Chapter 4. The location of the teletext data, and the special coding required for the MAC packet system, are functions of the MAC coder, in which these aspects form a background task. As the MAC systems have a great deal of flexibility from the data transmission point of view (conditional access, scrambling etc), the final details can be specified only as part of a complete system specification.

9 Teletext measurements

Video measurements

The teletext signal is carried by the normal programme video signal and therefore any teletext equipment such as data inserters, data bridges or regenerators which are in the main video signal path must have video performance reaching or exceeding normal broadcast standards so that the video signal is not perceptibly degraded. Furthermore, any crosstalk between the fast logic in the data processing circuits and the video channel itself must be at least 70 dB below the video signal levels to ensure that there is no possible interference in the video channel.

Typical performance figures and requirements for the inserter section of any teletext equipment are as follows:

1. Video input: 1 volt pk–pk into 75 ohms, with a picture to sync ratio of 70/30.
2. Return loss: better than 30 dB to 6 MHz.
3. Frequency response: ±0.05 dB to 6 MHz.
4. Luminance-to-chrominance delay: better than ±5 nanoseconds.
5. Pulse-to-bar ratio with 2T pulse: ±0.25 K.
6. Pulse shape performance (2T ±0.2 T pulse): ±0.25 K.
7. Bar overshoot: less than 1% of bar amplitude.
8. 50 Hz square wave tilt: less than 1% with respect to 0.7 volts.
9. Differential gain: 0.5% maximum.
10. Differential phase: 0.5 degrees maximum.
11. Non-linear distortions are to be measured with a test signal of three full bar lines and a five-step staircase waveform.

(The interested reader is referred to [20] for the principles of K-factor measurements.)

The inserter must also maintain similar performance when tested with a 'bump' waveform consisting of three lines of peak white alternating with three lines of black, followed by the staircase waveform.

The performance of the video clamps have to be such that the conversion noise does not degrade the noise performance below the required figure. Conversion noise is produced by the action of the clamps on signal noise and its effect is a low frequency variation of the line-to-line black level.

A most important requirement for any inserter is that it must not insert its signal onto an incorrect line, due for example, to some distortion in the input signal. This requires the sync separator to perform without error, in the presence of the types of distortions found in complex television networks. For example, on very long lines, very low frequency 'bounce' can cause the simple type of sync separator to produce spurious outputs.

Care must also be taken in installation lest reflections or echos occur due to poor terminations, distorting the digital signal and resulting in poor teletext performance.

Teletext data waveform measurements

For the maximum teletext service area from a transmitter the data must be transmitted to full specification. Normal video tests, however, do not check the quality of the teletext data signal and special measurements are therefore necessary. The position of the data signal on the video waveform, and its amplitude, can be measured by normal oscilloscope techniques using the white ITS bar as reference. If the data amplitude is low, reception of teletext will be impaired. If, on the other hand, the amplitude of the data is higher than specified then teletext reception might be improved but 'sound buzz' could be produced in the sound channel on certain receivers. Sound buzz is caused by high frequency video components of the data signal interfering with the inter-carrier sound IF signal. It is therefore essential that the transmitted data amplitude be correct. The data levels and timing are detailed in Figures 2.3 and 2.4.

The teletext data signal contained in the FBI is very difficult to measure accurately using a normal oscilloscope, particularly as the data is not locked to the video signal. To examine the data the oscilloscope time-base must be triggered from the data clock run-in at the start of the data line, but this requires a special trigger circuit sensitive only to the 6.9 MHz clock frequency. Examination of the 2 T pulse shape might provide some indication of the likely teletext performance but this cannot be relied upon. In general a poor group delay characteristic causes asymmetry and increased overshoots, as illustrated in Figure 9.1.

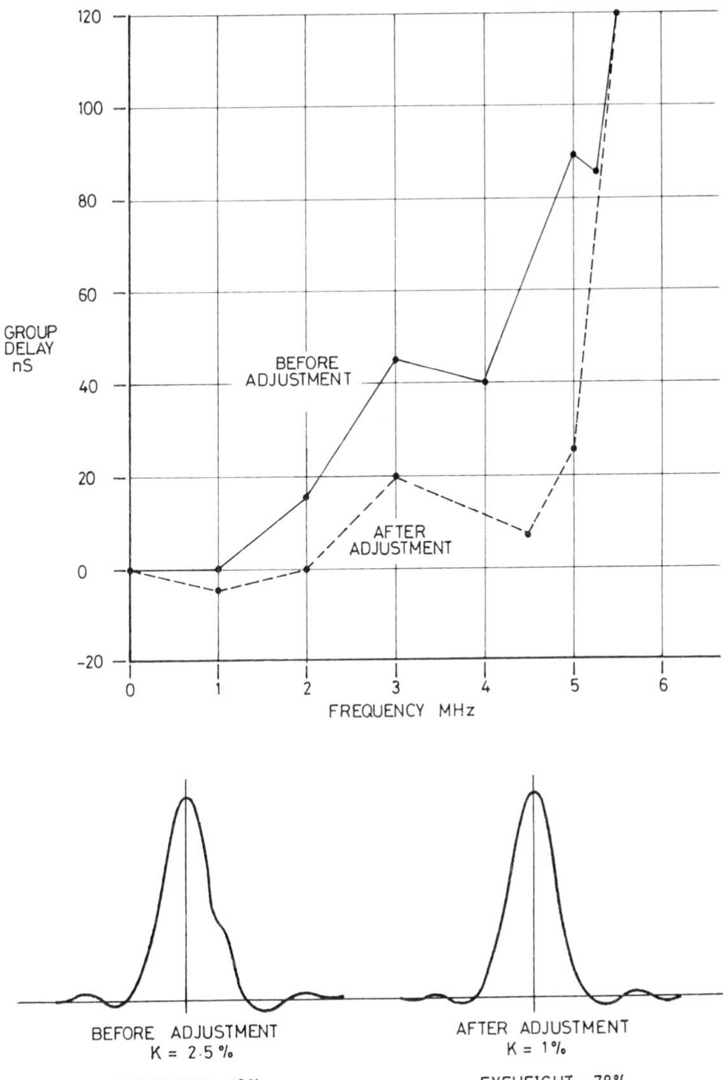

Figure 9.1 Transmitter phase response and 2T pulse shape

Some useful information about the data waveform can be obtained from a Lissajous figure displayed on an oscilloscope. The video signal is applied to the vertical deflection amplifier and a sub-multiple of the data clock (usually a quarter of clock

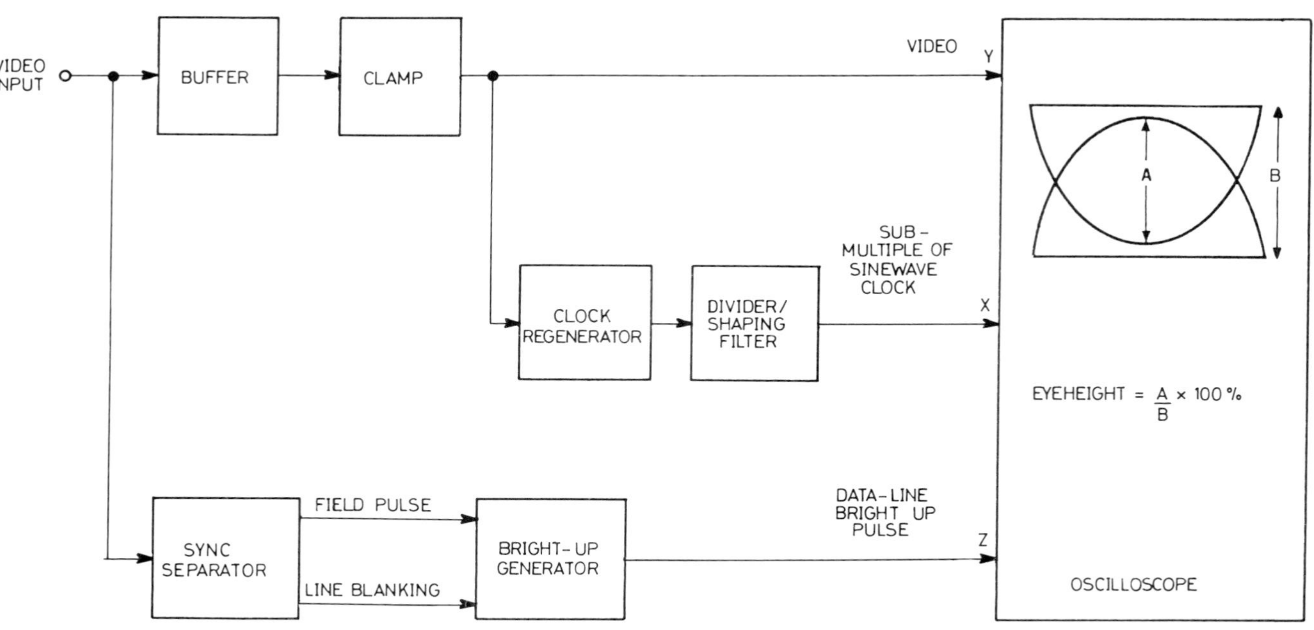

Figure 9.2 Lissajous eyeheight display facility

frequency) is used for for the horizontal deflection. A 'data-bright-up' pulse is also necessary to exclude all television information. Figure 9.2 illustrates the arrangement necessary to produce such a Lissajous display. The resulting display resembles an eye. The difference between the 0 and 1 levels is the 'eyeheight' and is normally expressed as a percentage of the true data amplitude. As the signal is degraded, the eyeheight falls. In the limit decoding becomes virtually impossible. Interpretation of the eye display is illustrated in Figure 9.3.

The internationally agreed definitions [4g] for the properties of the data waveform are as follows:

1. **All-noughts level**: the signal level resulting from a continuous stream of 0 pulses.
2. **All-ones level**: the signal level resulting from a continuous stream of 1 pulses.
3. **Midlevel**: the signal level corresponding to a level midway between the all-0 and the all-1 levels.
4. **Data amplitude**: a voltage corresponding to the difference between the all-0 and the all-1 level voltages.
5. **Noughts overshoots**: a voltage corresponding to the amount by which the peak value of signal voltage in the direction 1 to 0 exceeds the all 0-level signal voltage.
6. **Ones overshoots**: a voltage corresponding to the amount by which the peak value of the signal voltage in the direction 0 to 1 exceeds the all-1 level signal voltage.
7. **Peak-to-peak amplitude**: a voltage corresponding to the sum of the data amplitude, the 0 and the 1 overshoots.
8. **Eyeheight**: for a noise-free signal the eyeheight represents the smallest difference between any 1 pulse and any 0 pulse for sampling positions equally spaced at the data rate and positions chosen to maximize the quantity. It is expressed as a percentage of the data amplitude.
9. **Decision levels**: the level defining the decoding margin in the 1 region is the 1 decision level and the corresponding level in the 0 region is the 0 decision level.
10. **Decoding margin**: for a teletext signal referred to the clock run-in timing, the decoding margin represents the greatest difference between the 1 and the 0 decision levels for a given rate at which errors occur. The data samples are spaced at the data rate. It is expressed as a percentage of the normal data amplitude of the signal which should be measured in the absence of noise.
11. **Eyewidth**: for a noise-free teletext signal, the eyewidth

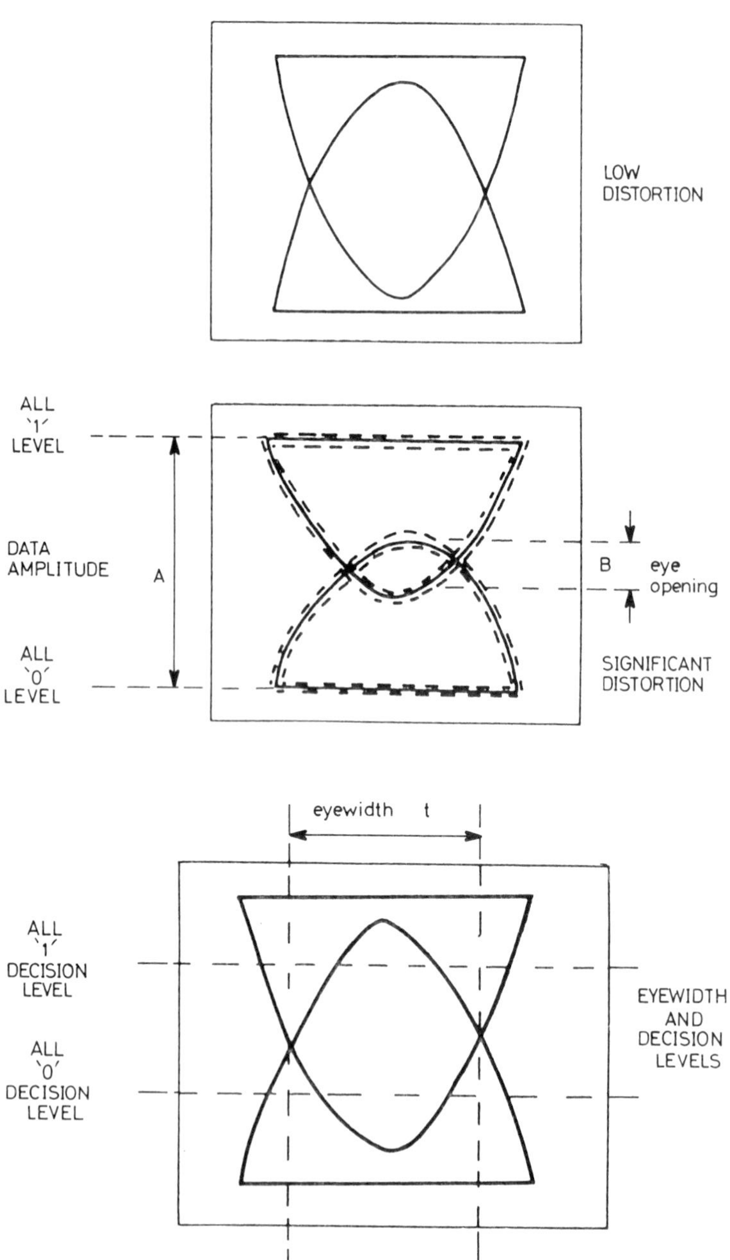

Figure 9.3 Eye patterns

represents the time interval over which error-free data results when the signal is compared with a given decision level chosen to maximize the quantity. It is expressed as a percentage of the bit period.

12. **Decoding threshold**: for equipment generating a graphic display from an applied teletext signal, the decoding threshold represents the smallest value of the decoding margin for a given character-failure rate of teletext signal composed of characters in a specified arrangement and degraded in a specified manner.
13. **Data asymmetry**: the amount by which the mean level of the eye pattern at the eyeheight measuring position on the time axis differs from the midlevel expressed as a percentage of the data amplitude.
14. **Noise interference**: the level of noise interference in this standard is defined as the RMS level of white wideband noise added to the baseband video signal before modulation, expressed in decibels, relative to a black-to-white level transition.
15. **Co-channel interference**: for performance testing of teletext equipment, co-channel interference may be simulated by the addition of a sine wave signal to the baseband video signal. The equivalent level of co-channel interference produced is taken as 11 dB greater than the RMS value of such a sine wave signal.
16. **Proportional jitter**: the percentage of the bit period not occupied by the eyewidth.

The Lissajous figure method of measurement suffers a fundamental disadvantage in that the reference data clock must be extracted from the distorted data stream. Any consequent phase jitter of the clock will cause phase jitter of the horizontal scan waveform and reduce the apparent height of the eye opening. This difficulty increases with low values of eyeheight as extraction of a jitter-free clock reference is then more difficult. Typical distorted eye displays are shown in Figure 9.4, and the difficulty of accurate measurement is obvious.

The data quality may be estimated by direct inspection, on a normal oscilloscope, of the data stream display of the clock cracker page, Figure 9.5. As can be seen, this test page is composed of two symbols only. The binary codes for these symbols (see code table, Figure 5.1) have the minimum number of transitions. This page therefore tests the decoder clock recovery circuit and also provides the least confused bit stream for visual

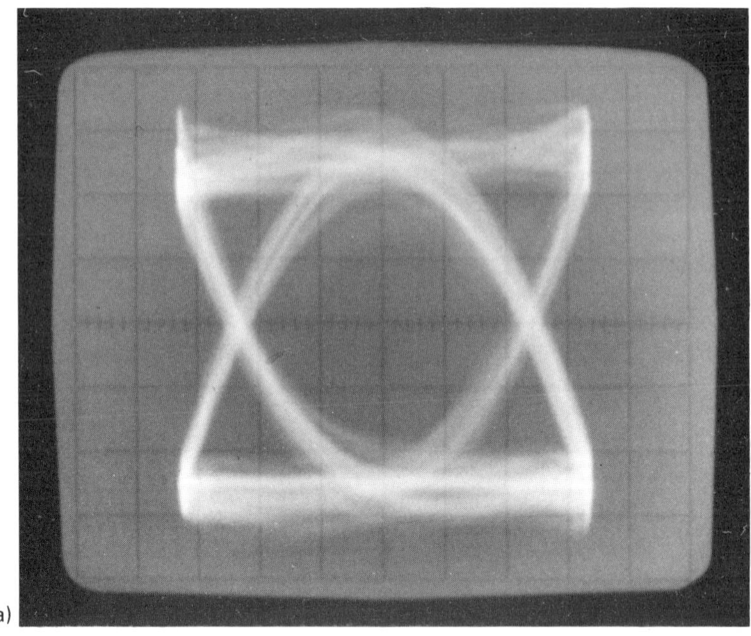

(a)

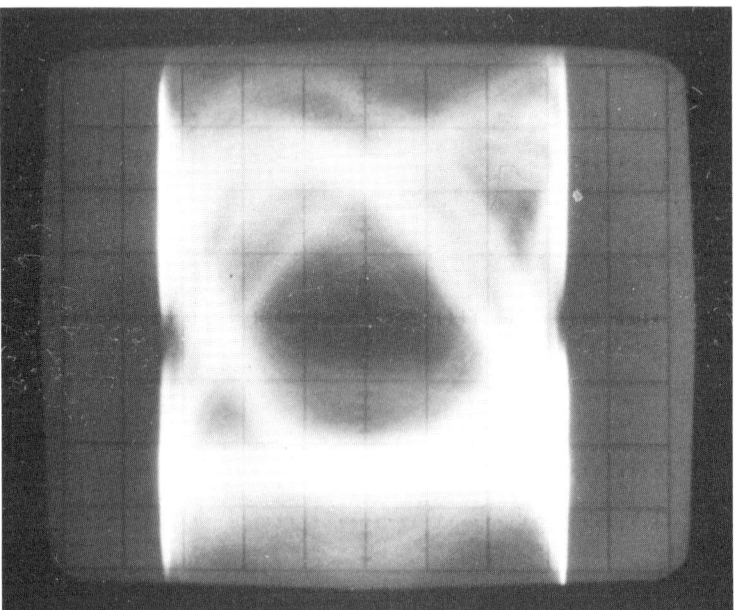

(b)

Figure 9.4 Distorted eye displays

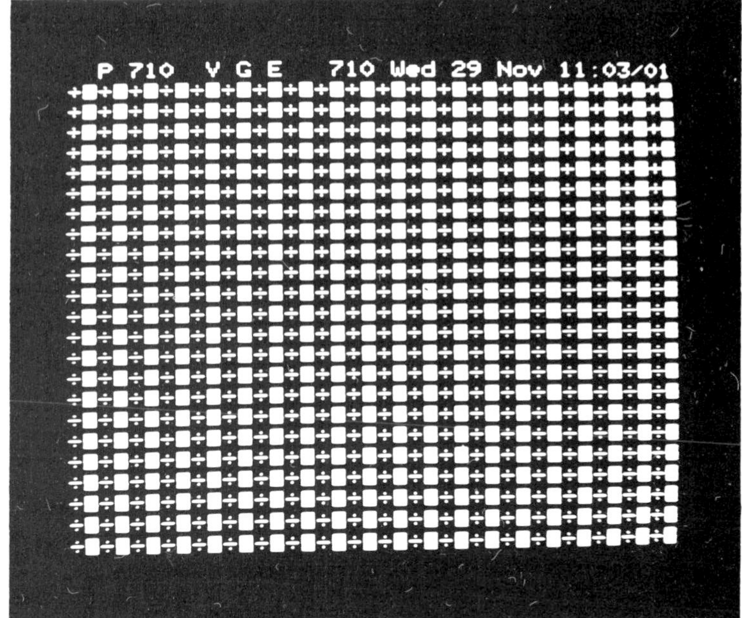

Figure 9.5 Clock cracker page

estimation of eyeheight or decoding margin. Errors on the displayed page of text are, of course, very easily seen.

Decoding margin

The most meaningful parameter of received teletext data quality is the decoding margin. This measurement takes into account the transmitted eyeheight, the noise and reflections which further degrade the data and jitter in the clock regenerator. Measurements made on an oscilloscope tend to be inaccurate since the parameters must often be estimated from very distorted eye displays, typical examples are shown in Figure 9.4.

To measure the decoding margin it is ncessary to determine the effective separation between the worst 0 and the worst 1 level over many data lines. A special instrument designed for this purpose by the BBC [21] is called a decoding margin meter, Figure 9.6. The instrument provides a digital display of decoding margin expressed as a percentage of the true data amplitude. The instrument is automatic in operation and a comparison technique is used so that the measurement is independent of data information or format.

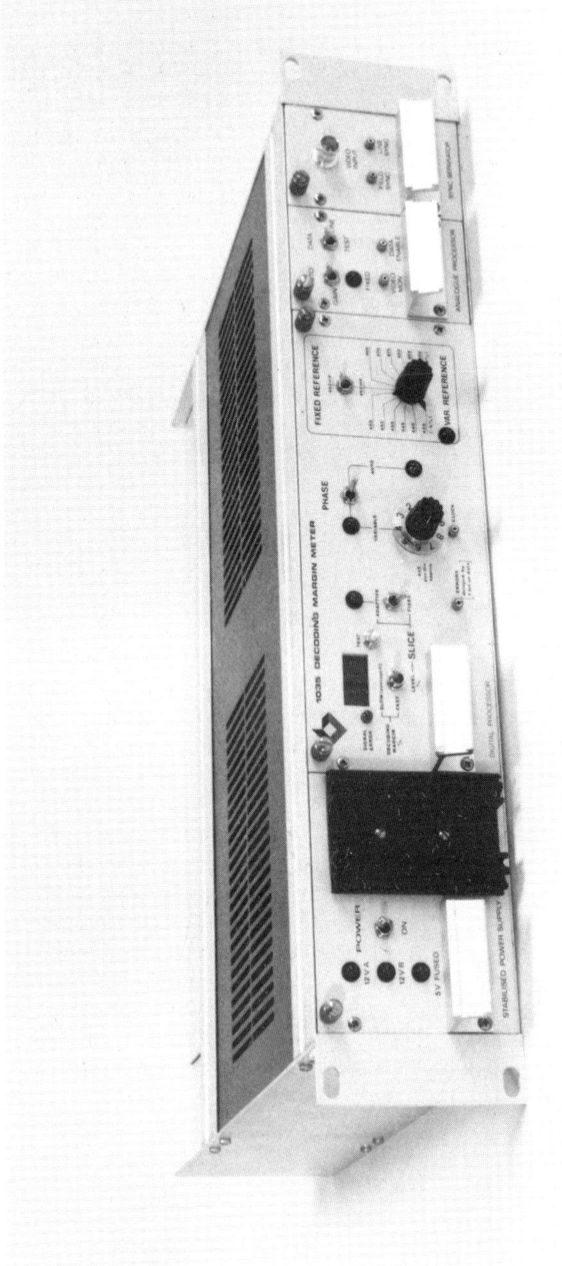

Figure 9.6 Decoding margin meter

A functional diagram of the instrument is shown in Figure 9.7. The input signal is automatically adjusted to a standard level by an AGC system using the VITS bar for reference so that the data amplitude at the inputs the measuring system is at is a standard value and, therefore, the measurement of the decoding margin is not dependent on amplitude. The clamped data signal is then applied to three slicing circuits operating in parallel. The output signals from the first and the third are compared with the output of the second slicer, which is used for reference purposes and has either a fixed or adaptive slicing level. The slicing levels of the first and third circuits are progressively changed, positively towards the 1 level and negatively towards the 0 level with respect to the reference slicing level. A difference signal generated by the comparitors indicates the level of the worst 0 or 1. The difference in potential between the slicing levels is then used for computing the decoding margin which is displayed as percentage of true data amplitude on a digital display.

The measurement can be averaged over 10 or 1000 data lines and hence takes into account the effects of noise and co-channel interference. A fixed or adaptive slicing level can be used for the reference slicer. Measurement of the adaptive slice level provides an indication of data symmetry (quadrature distortion). Consistent monitoring of the data signal quality is therefore possible and an output is provided for a remote display or for an automatic data logger. Also an alarm signal is generated if the data is lost. To assist in detailed analysis of distorted data using an oscilloscope bright-up pulses are generated, corresponding to the worst 0 and 1 in the bit stream.

Television signals are often distributed by widespread networks containing video links, main transmitters fed by re-broadcast links, and transposers [22]. Although equipment is maintained to video standards such networks can cause degradation of the teletext data signal. This is illustrated in Figure 9.8, which shows the improvements in radiated data quality (decoding margin) which results from data regeneration at the main transmitters (Chapter 6, page 65). For simplicity, the degradation in the decoding margin has been assumed to add arithmetically but in practice this is not the case and measurements must be made.

Decoder measurements

Decoders and equipment employing decoding circuits such as data bridges and regenerators must operate reliably even with distorted

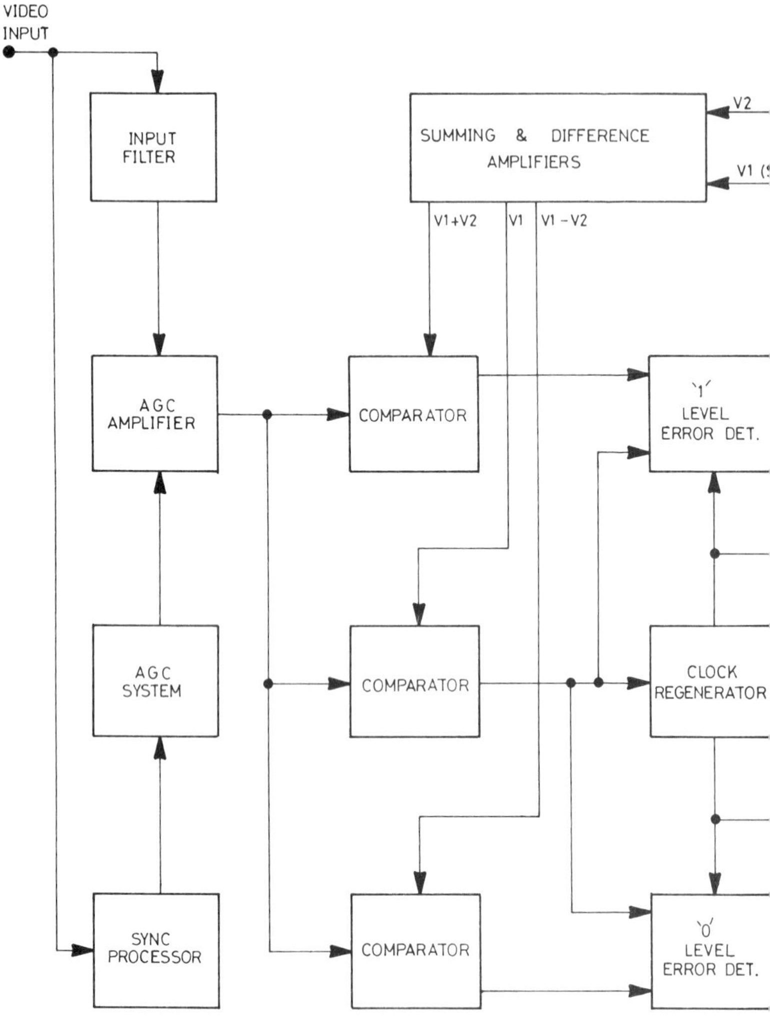

Figure 9.7 Functional diagram of an automatic decoding margin meter

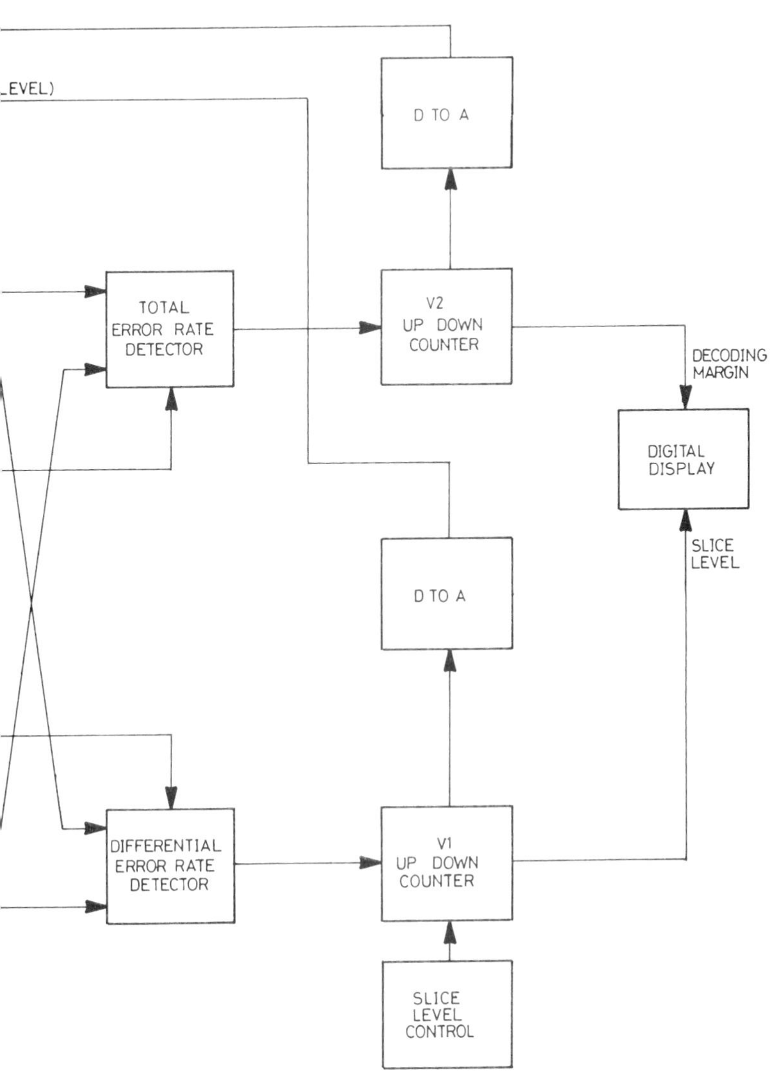

LEVEL)
D TO A
TOTAL
ERROR RATE
DETECTOR
V2
UP DOWN
COUNTER
DECODING
MARGIN
DIGITAL
DISPLAY
SLICE
LEVEL
D TO A
DIFFERENTIAL
ERROR RATE
DETECTOR
V1
UP DOWN
COUNTER
SLICE
LEVEL
CONTROL

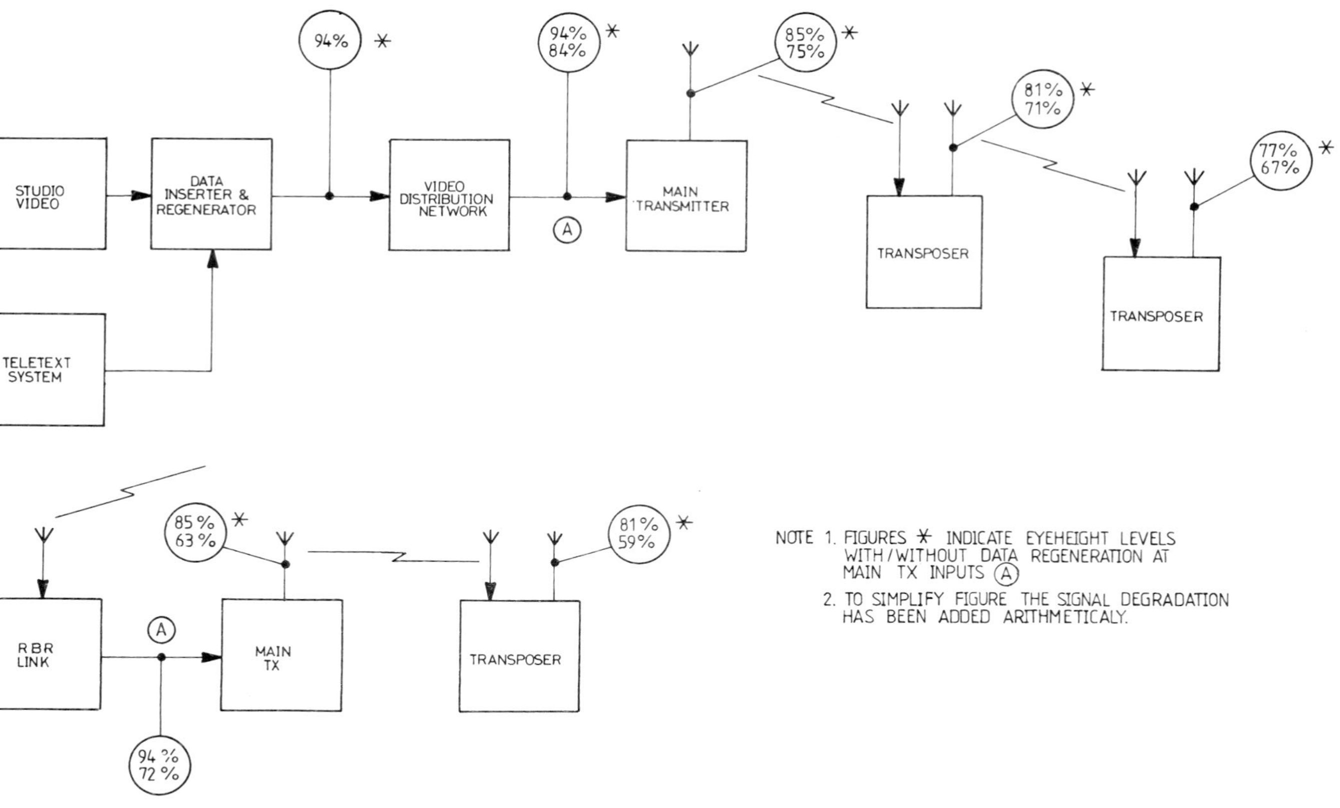

Figure 8.8 Degradation of data (eyeheight) in a broadcast network

input signals. To measure and accurately assess the performance of such equipment requires a source of signals with controlled levels of distortion. The design of teletext receivers for domestic use also requires such signals so that the performance of IF amplifiers, tuning systems and decoders can be assessed and compared. A source of teletext signals with various types of controlled distortion is therefore essential for design and product quality control purposes. The functional diagram of a calibrated distortion unit designed for this purpose is shown in Figure 9.9 and a photograph of the instrument is shown in Figure 9.10.

Referring to Figure 9.10, a teletext test page is generated to full broadcast specification and added to a locally generated composite sync signal. The data can be distorted independently in a number of ways. Firstly, the 'Delphi' principle (Defined Eye Loss with Precision Held Indication), a technique developed by the IBA [23], is used to generate reflections whch can then be added to the data signal. The reflection amplitude can be controlled between 0 and 100% in fixed or variable steps. Secondly, co-channel interference of two different frequencies can be added to the signal and, thirdly, white noise can be added. Both of these additional distortions can be separately added in controlled amounts.

The video signal itself represents a grating, with the vertical bar waveform corresponding to the shape of an elemental data pulse. The combination of the horizontal and vertical lines therefore produces a 'pulse-and-bar' waveform which allows the video characteristics of the receiver to be checked. A complete teletext page can thus be generated, with individually controlled levels of reflections (eyeheight), co-channel interference and white noise. This signal may be fed to a receiver, together with the video pulse-and-bar (grating) waveform. The signal can be fed either at video, direct to the teletext decoder, or, using an RF modulator, to the receiver aerial input. Very searching tests may be made of both decoder and receiver performance, and precise distortion levels at which received errors start to be produced in the page may be determined.

If a teletext page is received with errors on the first acquisition, the errors are normally corrected when the page is received a second time. However, to assess the performance of a decoder or receiver it is necessary to count the errors that occur on the first acquisition. The 'update' bit is therefore set in the header row of the transmitted page so that the memory is erased just before each new reception of the page.

The page repetition rate may be set to 1, 10 or 100 pages per

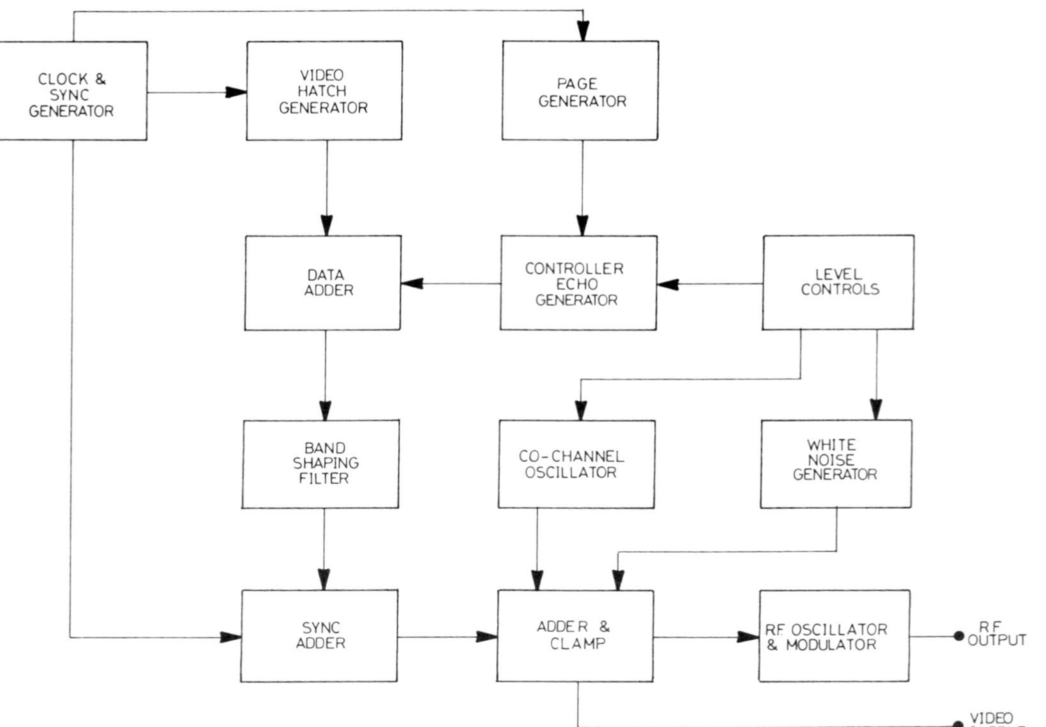

Figure 9.9 Teletext calibrated distortion unit

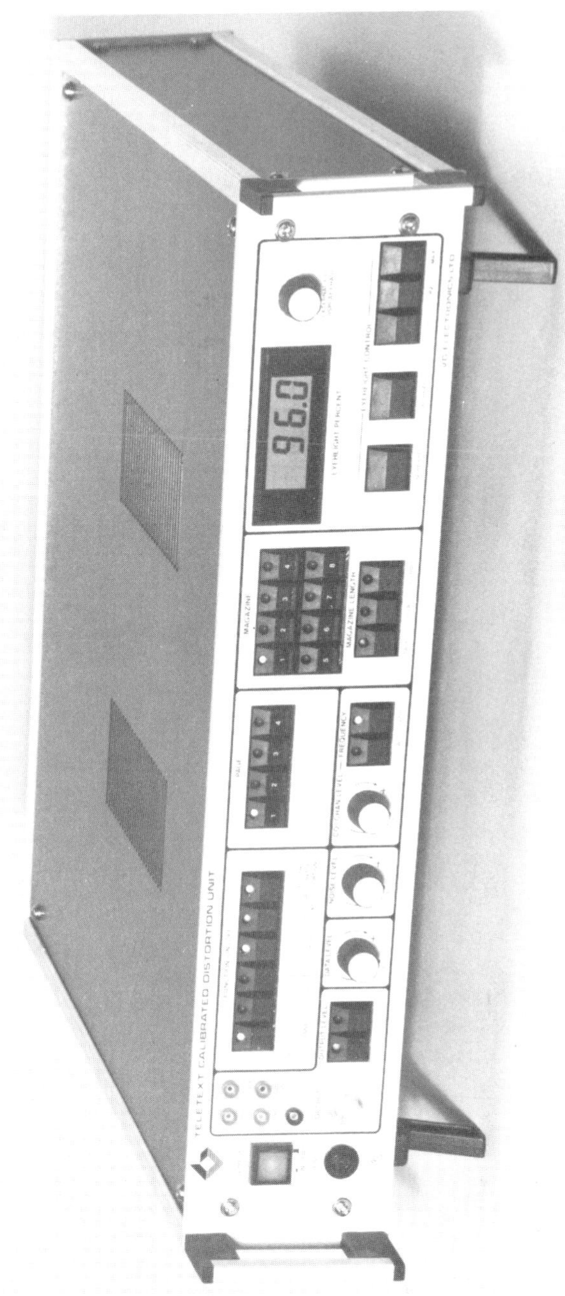

Figure 9.10 Calibrated distortion unit

second. The latter rate enables virtually a continuous page to be displayed, which shows up errors on a continuous basis. This is a useful technique when making adjustments to decoder clock circuits or receiver alignment. Alternatively, the slower rates give a longer time elapse between each page transmission which allows the particular errors on a page to be examined, or the mechanism which produces the errors to be further investigated, before the page is retransmitted.

The accepted design target for teletext receivers is for operation with input signals having decoding margins down to 25%, with two or three errors occurring when a page is first acquired. Allowing for degradation in the tuner and IF circuits, it follows that the decoder needs to operate with input signals having a decoding margin of about 20%.

When testing receiver performance in development or production it is also necessary to generate a magazine of special pages so that all the various aspects of a receiver's decoder and remote control system can be checked thoroughly and evaluated in a non-ambiguous manner. A teletext test page generator for this purpose must therefore meet the full broadcast specification and have appropriate pages programmed to meet the full broadcast requirements of the particular countries in which the receivers will eventually be used. It is particularly important to ensure that the transmission of 'multi-level' pages, that is, pages which contain additional data packets for language requirements, are configured to exactly the same specification as the final broadcasting system on which the receiver will be used.

Decoders for data reception

A teletext system can provide a data distribution service to specific groups of users, called 'closed user groups' [24]. The decoder for such applications might provide an RS232 output for driving a personal computer (PC), or the decoder might be contained in the PC itself. With a normal teletext transmission the magazine of pages is continuously repeated. If an error occurs in the reception of a page it is normally corrected when the page is next received. When a teletext system is used for data transmission this error correction arrangement cannot normally be used. A broadcaster will have several different closed user groups and furthermore the data will be continuously changing. It will therefore be transmitted only once or twice and it may well contain eight bit codes consisting of all 0 or 1 levels. The performance requirements of the decoder are therefore more onerous and a typical criterion would be that at least 100 megabytes of data must be received without error.

A decoder for data reception must first be checked as a normal teletext decoder using a calibrated distortion unit followed by checks for possible error rates. The error rate for data reception can be checked by using a teletext test generator producing a repeating data stream containing the full range of codes to be used, together with a cyclic redundancy check word (CRC). The RS232 output from the decoder is fed to a PC which contains special monitoring software. All errors in the received data and the continuity index are counted using the CRC check to confirm correct data reception. Details of all errors are displayed.

In data transmission systems for closed user groups the data is encrypted by the originator so that only receivers that are provided with the appropriate software can make use of the data, the teletext system itself being transparent. An alternative method of 'conditional access' can be used on an individual decoder basis. In such a system an individual user code has to be entered into the decoder to initiate reception. To allow a decoder to be tested without a particular 'user-code', an 'open' code can be used. If the decoder functions correctly with this code, no individual user-code tests are necessary. The test programme must be designed as part of the data management system.

The decoder forms only one component of a complete data transmission system. The decoder test progamme or the test generator can also be used at the data source to check the transmission system, including the distribution network and transmitters, for potential error-producing faults.

10 Radio Data System (RDS)

Introduction

Entertainment and education are the principal objectives of most television and radio broadcasting organizations. Various proposals have been made for broadcast stations to carry additional material to provide further usage of both the channel space and the transmission equipment. In the case of television signals, spare time exists in the FBI between each video field allowing data signals, such as teletext, to be added without causing any interference to the normal video or sound signals.

Analogue radio broadcasting signals do not have such a spare time slot and therefore any additional services require a sub-carrier located well above or below the audio signals. Data with very limited information rate can be transmitted as phase modulation of the carrier frequency. Such a system, called 'Teleswitching', is used in the UK, on the BBC longwave transmitter, for switching certain electrical power circuits a few times each day.

The modulation bandwidth of frequency modulated (FM) transmitters is typically about 90 kHz but the required bandwidth for stereo sound signals is some 53 kHz. Space is therefore available for information channels using sub-carriers well above 53 kHz. However, considerations of potential co-channel and adjacent channel interference limit the additional signals that can be accommodated. This is especially so in Europe where a very large number of FM stations operate in close proximity to each other.

For some years an information service concerning motoring conditions has been used in Germany. This operates with a suppressed carrier on a frequency of 57 kHz (three times the stereo tone pilot frequency). The service is called Autofahrer Rundfunk Information (ARI) and it is exclusively used for communicating motoring information for reception on car radio receivers which incorporate the necessary special decoder.

Over the period from about 1974 to 1982 a programme of work coordinated by the European Broadcasting Union (EBU) was undertaken by several broadcasting organizations. The objectives were to establish a data transmission system which would meet the stringent requirements placed upon it by the need for compatability and ruggedness when such a signal is carried on a sub-carrier alongside high quality stereo VHF–FM broadcasts. The laboratory work was backed up by over-the-air field tests conducted outside of programme hours, using a wide range of receivers. Once confidence had been gained in its compatability, the experimental system was put on-air from a number of transmitters in various European countries. These experimental transmissions, spanning several years, gave further confidence that the system had adequate compatability with the widest range of receivers and receiving conditions.

The final system having been designed, tested and agreed by the working group and formally endorsed by the EBU, Tech. Specification 3244-E was published by the EBU [25]. The system is known as the Radio Data System (RDS) and conforms with CCIR Recommendation 450/1. The EBU have now published Technical Document 3260, *Guidelines for the Implementation of the RDS system* [26]. Much of the development work undertaken during the development of the system is described; in addition, recommendations for practical implementation are provided.

The data sub-carrier

The frequency chosen for the data sub-carrier is 57 kHz, three times the frequency of the stereo pilot tone and synchronized in quadrature to its third harmonic during stereo broadcasts. During monophonic broadcasts the data sub-carrier is unsynchronized but retains the same close tolerance on its frequency as for stereo, that is ±6 Hz. This fixed relationship to the pilot tone makes an important contribution to compatibility, minimizing the audibility of any beats generated under multipath conditions, or in receivers that are not properly aligned. The sub-carrier is modulated by shaped biphase-coded symbols, representing the binary data stream. The modulation system used is two-phase phase-shift keying, with a phase deviation of 90 degrees. This system gives rise to a null at the sub-carrier frequency itself, with all the energy concentrated around a pair of side-bands separated from the centre frequency by the data rate. It is this spectrum-shaping which ensures compatibility with the German ARI system, which

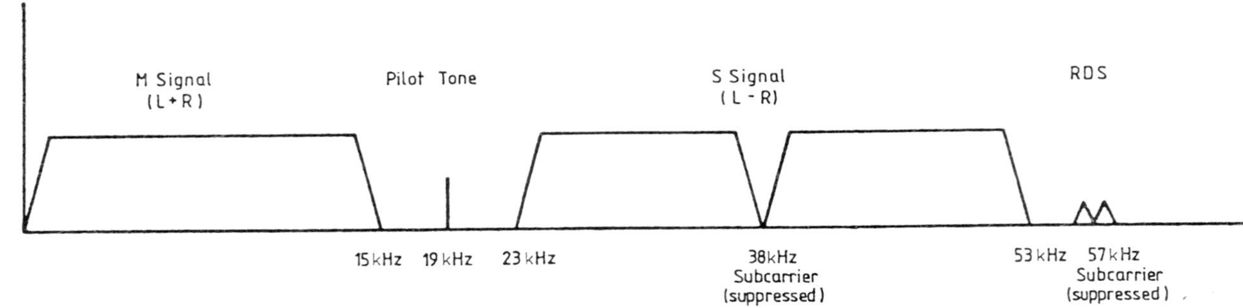

Figure 10.1 Stereo-sound MPX signal spectrum showing location of the RDS signal

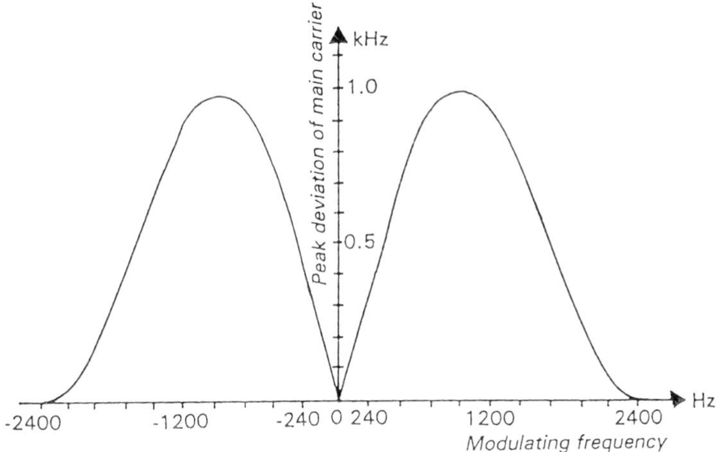

Figure 10.2 Spectrum of biphase coded radio data pulses

uses tones around 57 kHz. Figure 10.1 shows the baseband spectrum of the composite audio and RDS signal, and Figure 10.2 shows detail of the spectrum.

The amplitude of the data signal is also an important parameter. The larger the amplitude of the data signal the easier the reception; but, on the other hand, a larger data signal is liable to produce cross-talk or interference with the audio channel, particularly under non-ideal reception conditions. As a result of laboratory tests and on-air field trials, the data level is broadcast at 3% of the peak audio signal.

The data signal

The data rate of the signal used to modulate the sub-carrier is 1187.5 bits per second. This rate is derived from the sub-carrier frequency by division by 48. The phase locking of all the signals in this way leads to a simplification in the design of decoders. The data rate itself is constrained by considerations of compatibility and ruggedness, but the figure adopted gives adequate capacity for the various planned applications as well as allowing scope for future developments. The data is in two categories, fixed data that provides details of the particular transmitter frequency and station identity and possibly the frequencies of adjacent transmitters and variable data for different applications or services.

The RDS coder is normally situated at the transmitter and the variable data signals are fed to it over a telecom link. The fixed data is normally held in EPROM in the coder. A functional diagram of an RDS coder is shown in Figure 10.3. The various data entry ports for different services are connected to the system processor, which has its operating software held in EPROM.

The data from these external inputs is checked for errors in the usual way, and entered into the RAM memory. The processor selects the relevant data from the memories, calculates the CRC check word for transmission, where required, and the 'assembler' arranges the data into the specified format of data blocks for

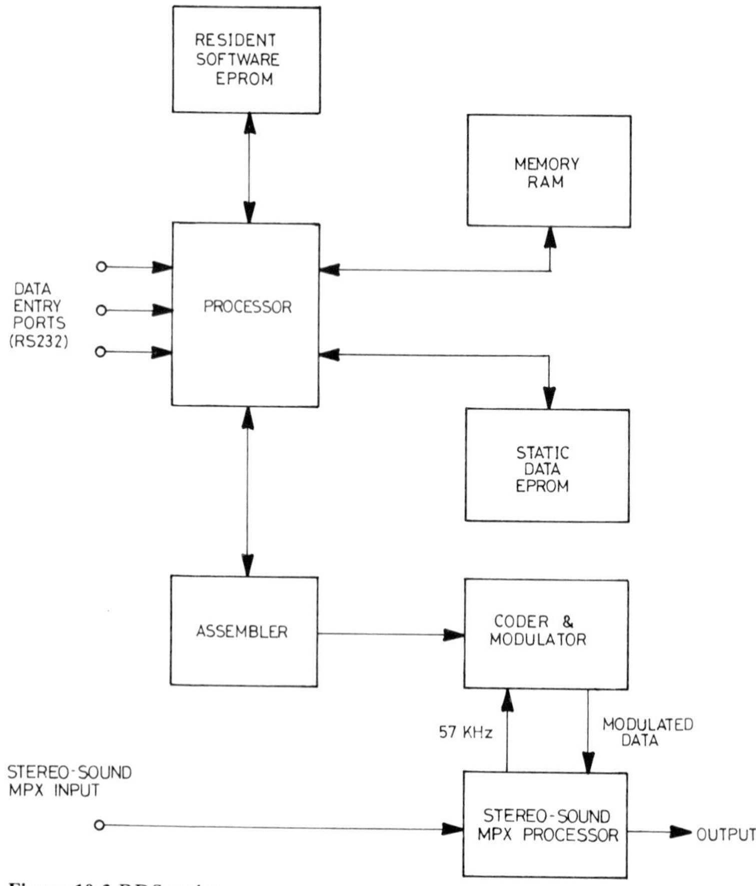

Figure 10.3 RDS coder

transmission. This data format is specified so that any RDS decoder can decode the data into a data block for processing according to the required application. In this respect the radio data system acts as a one-way modem. Update information can also be used to control the frequency and order in which the various fixed data groups are generated, and a simple command can be sent to the processor to initiate the actual transmission at the required time.

Referring again to Figure 10.3, the composite stereo sound multiplexed signal is applied to the MPX processor, which performs two functions. Firstly, it selects the 19 kHz pilot tone and locks the phase of the 57 kHz RDS carrier to it. This carrier signal is applied to the RDS modulator. The modulated RDS signal is applied to the MPX processor where it is added to the stereo-sound MPX signal at the appropriate level. The output from this unit drives the transmitter. Bi-phase coding is used for the data: a transition from positive to negative indicates a 1 and negative to positive a 0. The time-function for such data pulses is shown in Figure 10.4.

The general principle of RDS data coding is illustrated in Figure 10.5. The RDS carrier frequency of 57 kHz is first divided by 24 to produce a signal at 2375 Hz, and then divided by two to produce the data clock at 1187.5 Hz.

When the data signal is demodulated in a receiver it may be inverted (unlike television, where the polarity is determined by the video signal), and differential coding is used so that the data is correct irrespective of polarity. The data is coded according to the expression:

Input (t)	+ input	($t - t_d$)	= Output (t)
0	+	0	= 0
1	+	0	= 1
0	+	1	= 1
1	+	1	= 0

where t_d = 1 bit period.

When the input is 0 the output remains unchanged from the previous output, but when an input 1 occurs, the new output bit is the complement of the previous bit.

Differentially coded data (waveform 2, Figure 10.5), is produced by feeding the input data to an exclusive OR gate, the other input being the input data delayed by one bit period. The NRZ symbol generator produces positive pulses for each bit period when the input is high and negative pulses when input is

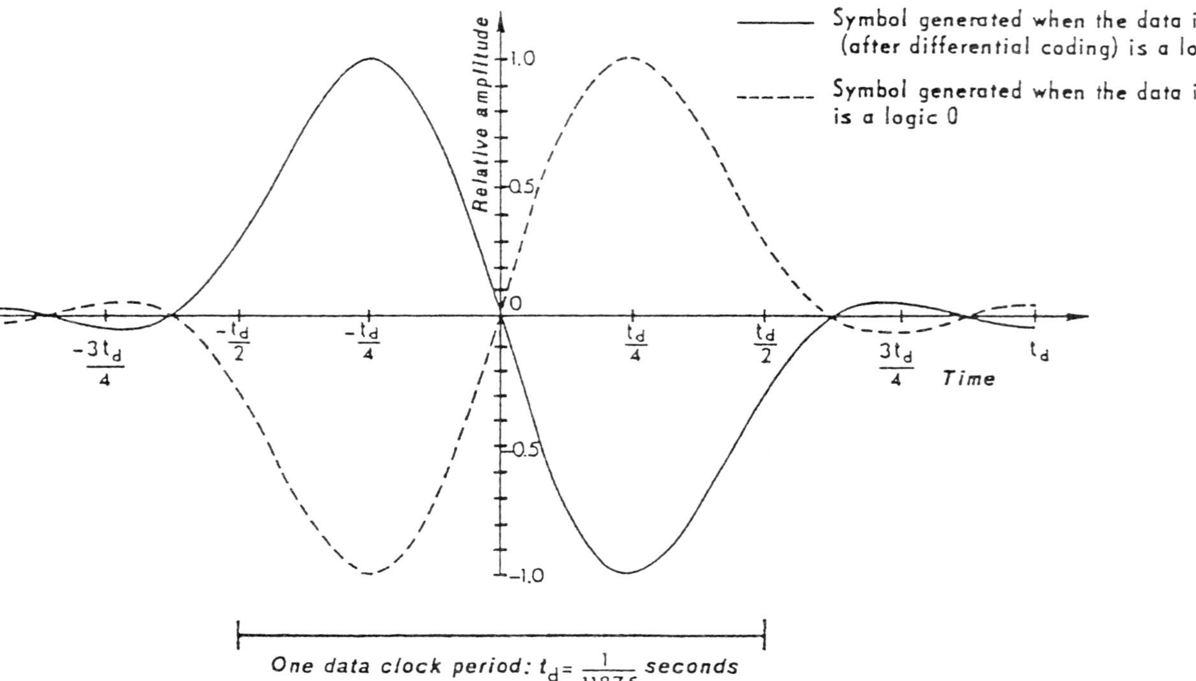

Figure 10.4 Time-function of a single biphase symbol

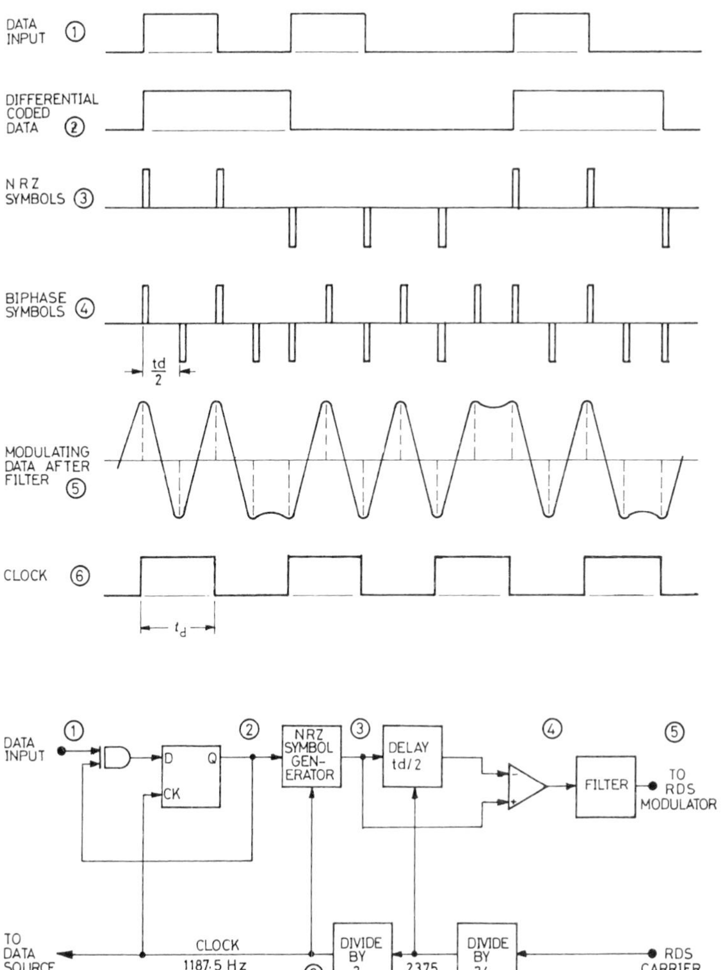

Figure 10.5 Principle of RDS data coding

low (waveform 3). The bi-phase symbol generator produces a negative going pulse $t_d/2$ after each positive pulse and a positive pulse after every negative pulse (waveform 4). After passing through the shaping filter, which has a bandwidth of approximately 5 kHz, centred on 57 kHz, the RDS modulation (waveform 5) is obtained. This signal is then added to the multiplexed stereo signal at an amplitude which is approximately 3% of that due to the

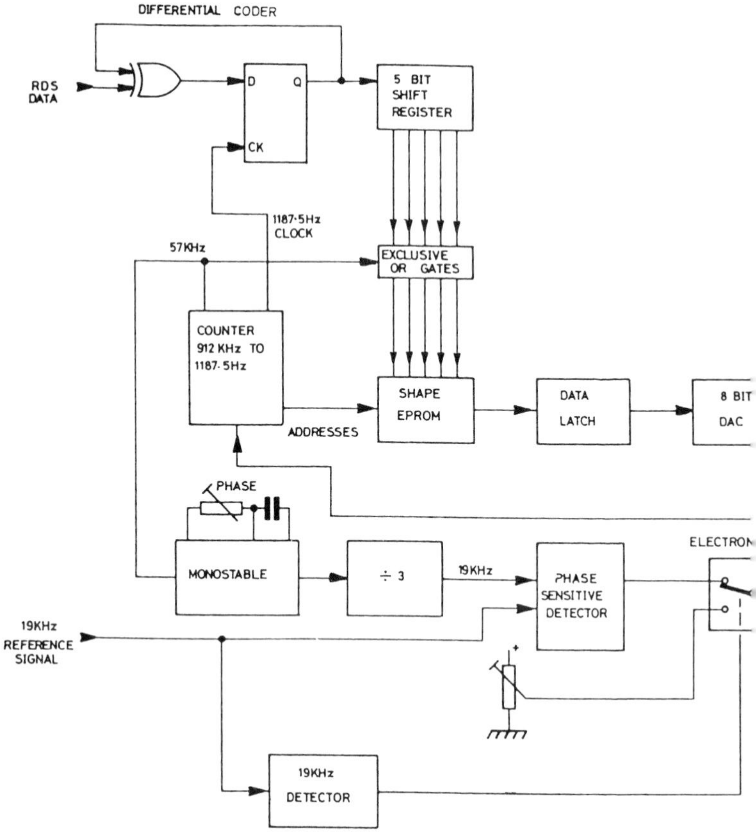

Figure 10.6 RDS coder with digital filter

sound signals. Commercial RDS coders usually use a digital filter to obtain optimum pulse shape.

The functional diagram of a coder based on a BBC design is shown in Figure 10.6. The master clock oscillator frequency is 3.648 MHz and is phase-locked to the third harmonic of the 19 kHz pilot tone. It is divided by four to generate a 912 kHz clock for digital generation of the data signal. This clock is further divided by 16 to produce the 57 kHz carrier, by 48 to produce a clock at 3.562 kHz, and by three to produce the bit rate clock at 1187.5 Hz.

The input data is clocked and differentially coded to feed a 5-bit shift register. The shift register makes available, relative to a 'middle' bit, the two preceding bits and the next two bits to be transmitted. These are needed to generate the intersymbol

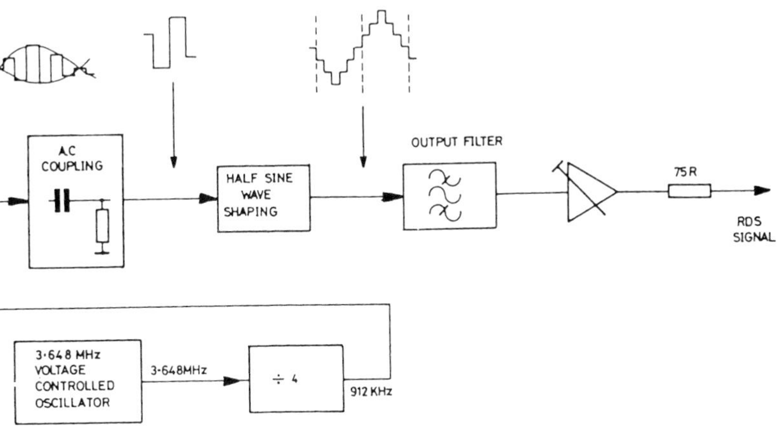

components of the data waveform. These five bits are fed to an EPROM containing a 'look-up' table corresponding to the RDS data shape. This data is clocked out using an address generated from the internal clock locked to a multiple of the RDS data rate. The modulation process to form the final suppressed carrier signal is achieved by inverting the five data bits fed to the EPROM once every half cycle of 57 kHz. This changes the output of the EPROM from a simple RDS wave shape to a suppressed carrier double side-band signal, based on 57 kHz. The output is passed through a latch, to remove 'glitches', and then fed to a digital-to-analogue converter. The data waveform has now become the envelope of a 57 kHz suppressed carrier signal but is rich in harmonics. It is further shaped by imposing a half sine-wave shape on each half cycle. An eight-input analogue multiplexer selects the correct attenuation for the rectangular data signal eight times every half cycle of the 57 kHz signal. This signal is then buffered and band filtered to produce the RDS output signal.

Decoding the RDS signal

The input signal for the decoder is taken from the demodulator prior to any de-emphasis filtering and is band-filtered to separate the 57 kHz RDS signal. This signal is synchronously demodulated using a locally generated 57 kHz sub-carrier. A 2.4 kHz low pass filter is then used to separate the RDS modulation.

The principle of decoding RDS data is illustrated in Figure 10.7. The reference clock is regenerated from the data signal, or can be obtained by division of the 57 kHz carrier used to drive the synchronous demodulator. The phase of the reference clock is important as the transitions must coincide with the cross-over points of the RDS data waveform as illustrated in waveforms 1 and 2. (In the coder the clock is in phase with the peaks of the modulation.)

A switching circuit, driven by the regenerated clock, is used to invert every half cycle of the RDS waveform that corresponds to the positive clock period, as illustrated, waveform 3. A logic '1' symbol at the input corresponds to two positive half cycles at the output of the inverter (see Figure 10.4). The differentially coded data could be obtained from this waveform directly but the signal-to-noise ratio is significantly improved by using an integration circuit. The integrating circuit is switched by the regenerated clock so that after each complete clock cycle the output is returned to zero and the potential at the end of the integration cycle is stored. The actual potential reached at the end of each integration cycle will vary and therefore a slicing circuit is used to produce constant amplitude pulses as illustrated by waveform 5. A flip-flop circuit which changes state only when the pulses change polarity is used to produce the differentially coded data (waveform 6).

The regenerated clock could be 180 degrees out-of-phase which would produce an inverted data signal at this point (waveform 6). The output data is produced by the differential decoding circuit, as shown in Figure 10.7, the output data being independent of the clock polarity.

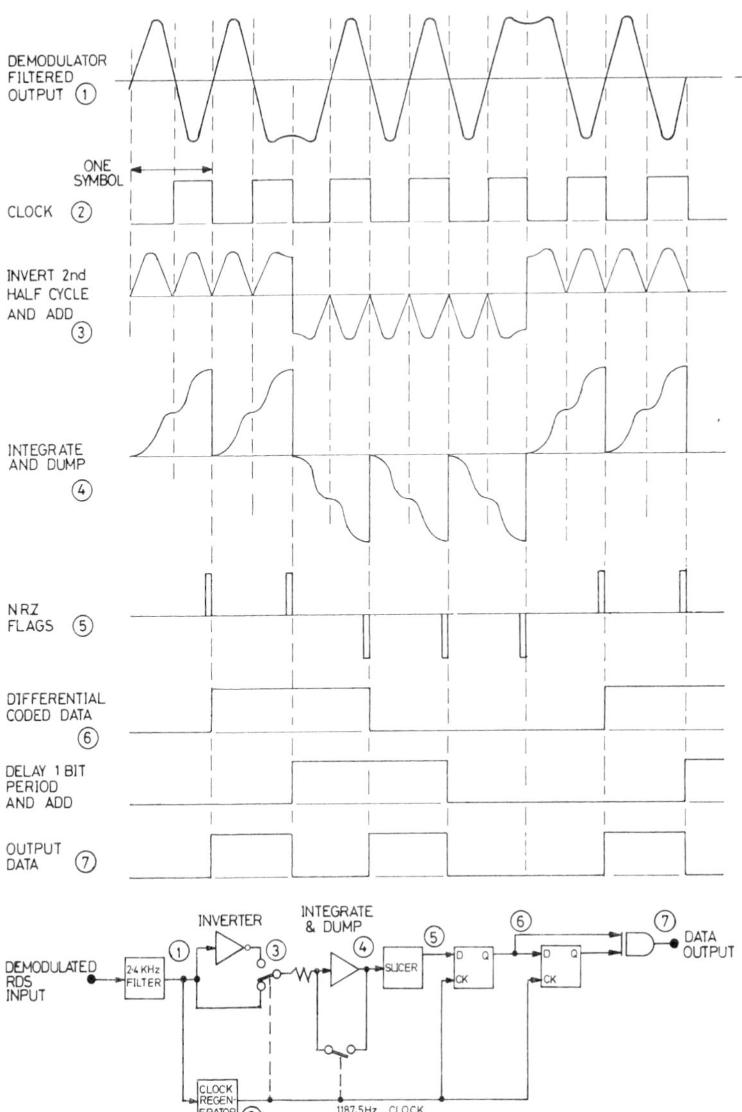

Figure 10.7 Principle of decoding RDS data

Decoding is according to the expression:

Input (t)	+ input	$(t - t_d)$	= Output (t)
(0)			
0	+	0	= 0
1	+	0	= 1
1	+	1	= 0
0	+	1	= 1

or if inverted

(1)			
1	+	1	= 0
0	+	1	= 1
0	+	0	= 0
1	+	0	= 1

where $t_d = 1$ bit period.

The decoded data and clock signals are fed to the data processing circuits in the receiver, or via a suitable drive circuit to an external PC.

Commercial receivers use hybrid decoding circuits which contain the 57 kHz band-pass filter, synchronous demodulator, bi-phase symbol generator and differential decoder as a single element. The multiplex signal derived from the demodulator is normally applied directly to the hybrid circuit and the raw data and clock are generated at logic levels, so they can be applied directly to a suitable processor circuit. The only external components required are the crystal, typically 4.332 MHz, for the phase-locked loop. This frequency is 76 times that of the RDS sub-carrier.

A decoder often forms part of an RDS coder at the transmitter. The transmitted data can then be checked automatically by comparison with data in the assembler. If an error is detected it can cause the data to be retransmitted or raise an alarm. It also allows RDS data received from a neighbouring transmitter to be decoded and used as an update source for the RDS coder.

11 RDS data format and applications

Data format

The EBU Specification 3244-E [24] defines the format of the transmitted RDS data, together with various applications which meet the requirements of several broadcasting organizations. Since their RDS transmissions conform to the specification any receiver manufacturer may produce receivers that will function correctly.

Broadly speaking, there are four main spheres of application currently envisaged.

1. Car radios, where auto-tuning is a most important benefit as it allows listeners to be mobile without the necessity of re-tuning.
2. Domestic radio receivers, on which the station name may be displayed to simplify tuning; and also clock time.
3. Radio paging, as pocket receivers can be addressed individually and are able to display telephone numbers or simple messages.
4. Radio text systems, allowing messages or information to be displayed on small LCD displays.

In the first three applications the receivers will be fully integrated, in that the RDS decoder, the associated processor and resident software will all be fitted when the receiver is manufactured. For the message system the RDS receiver may well be fitted with an RS232 output so that it can be linked to a PC and function as a one-way modem in a business system. To provide for future expansion, there are a number of application areas as yet unspecified but these can be simply implemented by transmitting appropriate designation codes as detailed in Specification 3244-E.

The data is transmitted as 16-bit words, each of which is associated with a special 10-bit check word for error protection. Four such 26-bit blocks form a ‘group’ as shown in Figure 11.1. Sixteen groups, each of which can be an A or B type, are allocated to cover the various applications, but to date only 11 have been specified, as indicated in Table 11.1. This allows for future

Table 11.1 EBU specified group types and applications

Group types (Bit 5 defines A or B)

	Code					*Application*
No.	*A3*	*A2*	*A1*	*A0*	*1*	
0	0	0	0	0	0/1	Station tuning and switching information
1	0	0	0	1	0/1	Programme item number (PIN)
2	0	0	1	0	0/1	Radiotext (RT)
3*	0	0	1	1	0/1	Other network information (ON)
4	0	1	0	0	0	Clock time and date (CT)
5	0	1	0	1	0/1	Transparent data channel for text (TDC)
6	0	1	1	0	0/1	In house application (IH)
7	0	1	1	1	1	Paging (PG)
8	1	0	0	0		Traffic Message Channel (TMC)
9 to 13	1 1	0 1	0 0	1 1		Not yet specified
14*	1	1	1	0	0/1	Enhanced other network information (EON)
15	1	1	1	1	1	Fast station tuning and switching information

* Group 3 (ON) and 14 (EON) must never be transmitted together.

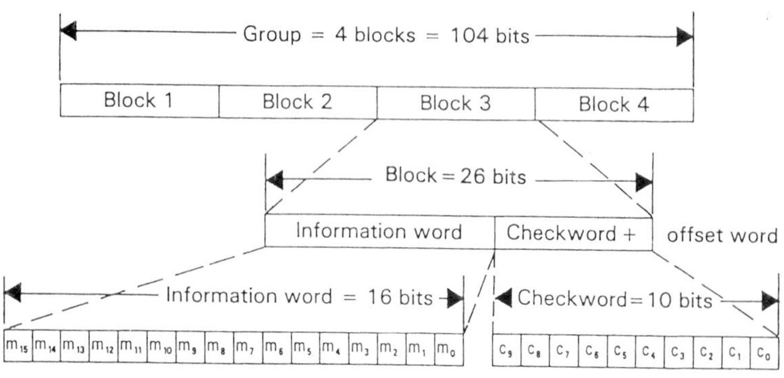

Figure 11.1 RDS group composition

developments. No additional bits are required for framing or other synchronization purposes as this information can be derived from the data stream itself, by the use of the special checkwords.

An important feature of the RDS system is its flexibility. The different group types can be inserted in any order, to suit the requirements of the particular set of applications chosen by the broadcaster at any given time. Each group is an entity in itself and can be decoded without reference to any other group. The only overriding requirement is that the RDS data block containing the programme identification (PI) code must occupy the same fixed position in every group and be repeated sufficiently frequently to allow receivers with automatic tuning to operate, with a reasonable response time.

The main features of the message structure are shown in Figure 11.2. The key functions are as follows:

The first data block in every group always starts with the PI code.

The first four bits of the second block of every group specify the application of the group. Groups are referred to as types 0 to 15. Each type has two versions (A or B) defined by the fifth bit of the block. In A versions the PI code is in block 1 only and in B versions it is in both blocks 1 and 3.

The programme type code (PTY) and traffic programme identification (TP) occupy fixed positions in block 2 of every group.

The PI, PTY and TP codes can be decoded independently of any other block to minimize the acquisition time for this type of message and also to retain the advantages of the short block length (26 bits). This is achieved by inserting a special word, 'offset C', in block 3 of version B groups. The presence of this word indicates to the decoder, without reference to the fifth bit of the application reference code group, that block 3 is a PI code. Table 11.2 gives the recommended minimum repetition rates given in the EBU specification for some of the main applications, and Table 11.3 a typical group mixture to achieve it.

RDS applications

Automatic station tuning

Several potential applications for RDS have been specified in detail; most are termed 'optional'. The 'compulsory' signals, which must be included in order to comply with the EBU specification, are associated with the identification of the radio station itself.

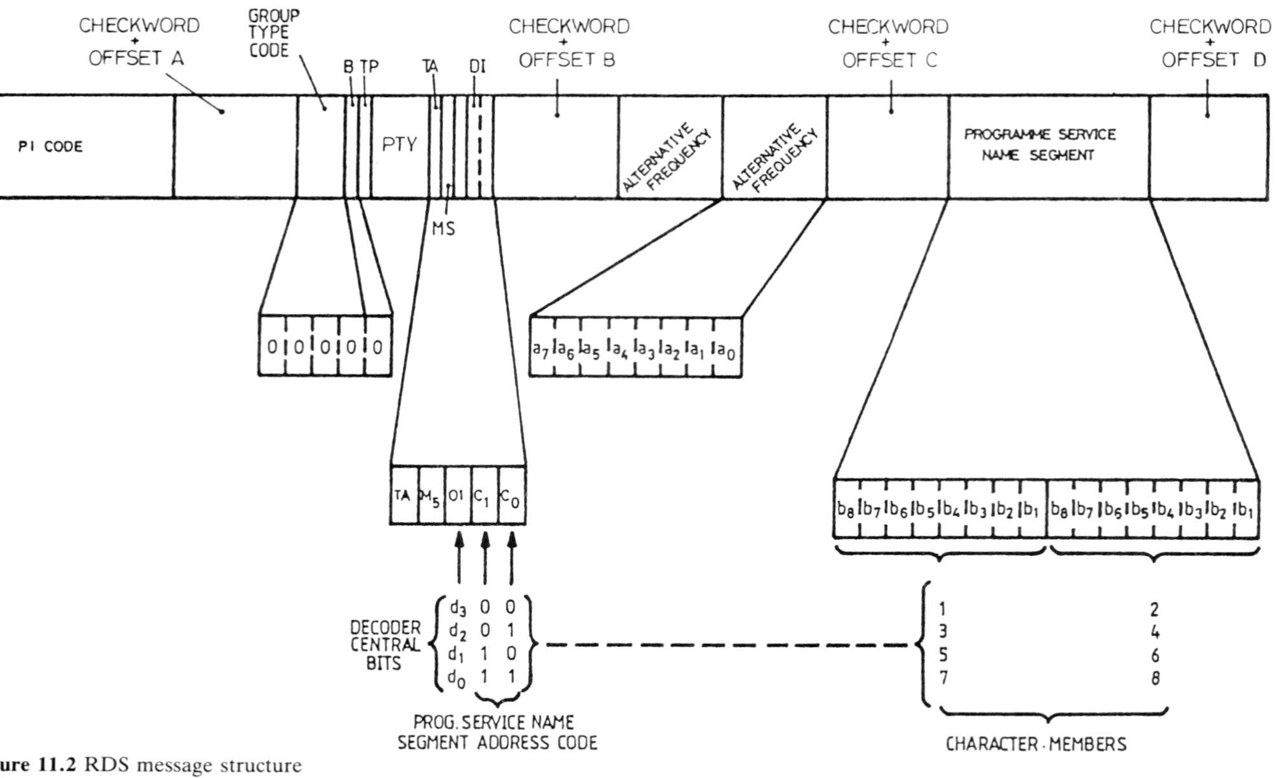

Figure 11.2 RDS message structure

Table 11.2 Recommended minimum repetition rates for some main applications (EBU Tech 3244-E)

Applications	*Group types which contain this information*	*Recommended minimum repetition rate per second*
Programme identification (PI) code	all	11
Programme service (PS) name	0A, 0B	1
Programme type (PTY) code	all	11
Traffic programme (TP) identification code	all	11
Alternative frequency (AF) code	0A	4
Traffic announcement (TA) code	0A, 0B, 15B	4
Decoder identification (DI) code	0A, 0B, 15B	1
Music speech (M/S) code	0A, 0B, 15B	4
Programme item number (PIN) code	1A, 1B	1
Radiotext (RT) message	2A, 2B	0.2

Table 11.3 Typical group mixture to achieve the recommended repetition rates

Group type	*Applications*	*Typical proportion of groups of this type transmitted*
0A or 0B	PI, PS, PTY, TP, AF, TA, DI, M/S	40%
1A or 1B	PI, PTY, TP, PIN	10%
2A or 2B	PI, PTY, TP, RT	15%
3A or 3B	PI, PTY, TP, ON	10%
Any	Optional applications	25%

These signals alone would allow the implementation of a simple 'automatic' receiver, with display of the station name.

The programme service name (PS) is a string of eight ASCII-coded alphanumeric characters intended for display on a receiver. The simplest possible RDS receiver would make use only of this feature, being otherwise tuned conventionally, but with a positive display of the identity of each station.

The PI is a 16-bit code to identify the particular radio station, or network, originating the broadcast. All transmitters carrying the same PI code carry the same sound signals. A receiver can therefore search for a particular PI code and thus find a particular broadcast service. The first four bits of the code identify the country of origin, the second four the type of service, (local, national etc.) and the final eight form a serial number for the particular station or network.

The PS and PI features alone would allow the implementation of receivers which could find and display a station's identity automatically but the service can be considerably enhanced if further information is transmitted. For example, consider the case of a car radio as it nears the edge of the service area of a particular network transmitter. As the signal fades, the receiver will recognize that it must find another transmitter carrying the same programme. It could do this by scanning the band, searching for the same PI code, but this would inevitably take a few seconds, particularly as the receiver would need to stop on each receivable signal for long enough to decode the RDS bit stream. A receiver with two 'front ends' could be doing this all the time, and loading its memory with details of receivable signals as it encountered them, but the whole process can be greatly simplified and speeded up if each transmitter radiates information concerning the frequencies on which its neighbours can be found. The receiver then need only look at these specific frequencies to check for a signal at a receivable level. It is anticipated that a receiver with a single front-end will be able to achieve this quite adequately.

The RDS feature which accommodates this type of information is called alternative frequencies (AF). A list of up to 25 frequencies can be transmitted, and it is important to note that MF and LF channels can be included, thus catering for the situation where VHF–FM coverage may be incomplete.

A further improvement to the potential utility of an RDS receiver is provided by the other networks feature (ON). Consider the case of a receiver which is in use on a particular station, and the listener wishes to quickly tune to an alternative station. With the information so far available, the receiver would again have to scan the band from one end to the other in order to identify the strongest signal which carried the appropriate PI code. The time taken could well be too long.

This problem has been overcome by the introduction of an improved arrangement called 'enhanced other networks' information (EON). Transmitted together each station's signal is information about frequencies on which other services, referenced by their PI codes, can be found. This improved arrangement gives the receiver a sequence of PI codes and associated frequencies, for each of the other services available in the service area of the transmitter. An RDS receiver therefore has available in memory the station frequencies for the alternative stations.

Broadcasting these four types of information allows the development of truly intelligent receivers which not only find a station on request but can also tune to the transmitter for best reception as the receiver moves about the country. The receiver can also respond almost instantaneously to a request for a different station. This is achieved with no demands on the user for pre-programming, or indeed for any knowledge of the station frequencies involved.

The functional diagram of such a receiver is shown in Figure 11.3. The user selects the programme and the processor controls the tuning so that the appropriate transmission is received. The RDS data is decoded and all the relevant information concerning alternative frequencies is stored in memory. The station name and clock time will also be displayed. Multipath reception causes a low frequency AM component to be produced by the AGC system. The AGC system can provide a signal representative of signal quality, and this may be used to initiate the selection of an alternative transmitter.

Traffic information

Traffic programme identification (TP) is an on/off switching signal to indicate to the receiver whether this is a programme which carries announcements for motorists.

Traffic announcement identification (TA) is also an on/off switching signal which indicates when a traffic announcement is actually on the air, thus allowing automatic switching away from another radio station, or from tape cassette listening.

The operation of auto-tuning and a traffic information service is shown in Figure 11.4, which illustrates reception in a motor car moving along a motorway. At the start of the journey the vehicle would normally receive signals from the main transmitter (A). This transmitter carries in its RDS data the frequencies of all the other transmitters located at the same site together with those in the neighbouring sites (EON). Traffic announcements for the local area are radiated by transmitter (B). In response to a signal from transmitter (B), a traffic announcement (TA) flag is include in the RDS data of the main transmitter located at (A). This causes the

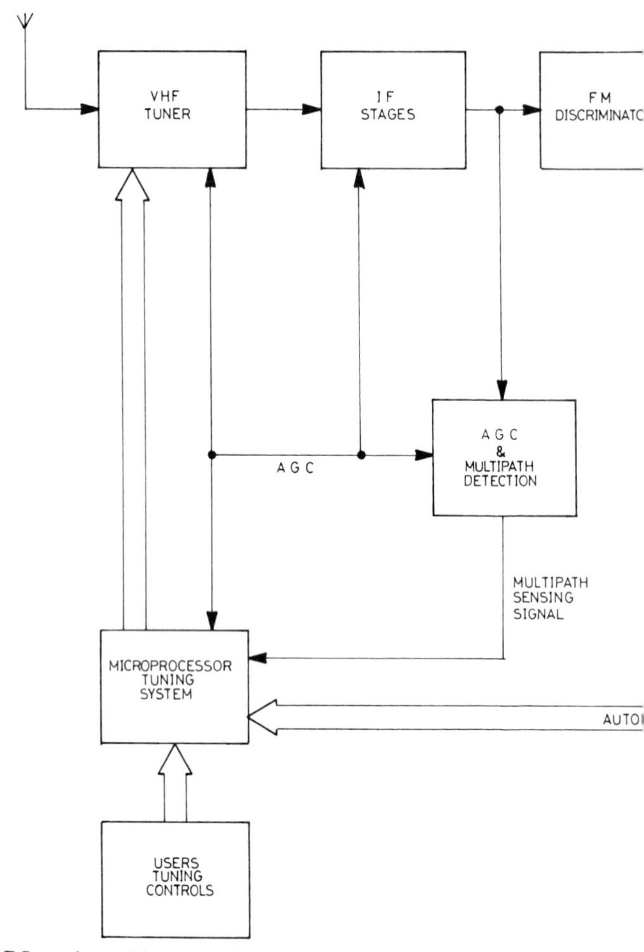

Figure 11.3 RDS receiver with automatic tuning

receiver in the car immediately to change channel to that of the local transmitter (B) on which the traffic announcement is about to be broadcast. The TA flag could, in a suitably designed receiver, cause the cassette recorder (if in use) to stop, so that the radio would automatically receive the traffic announcement from station (B). At the end of the announcement the TA flag would be cancelled in the RDS data from transmitter (B) so that the receiver immediately reverted to the original programme from the main transmitter, or the cassette would restart.

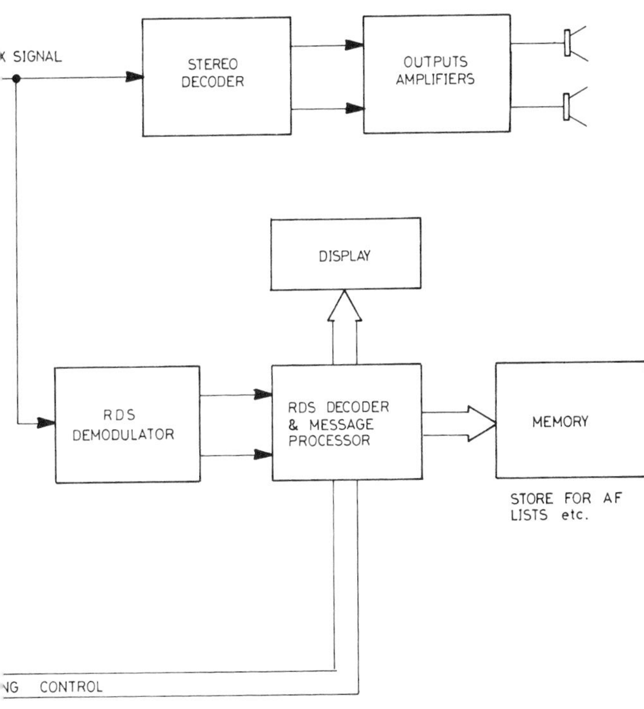

This arrangement allows local traffic announcements, which are relevant to a particular location, and broadcast by a local station, to interrupt a national programme (or tape). It frees the national networks from the need to broadcast such information and provides the motorist with immediate information on local conditions. As the car continues along the motorway, it will reach the fringe of the service area of the main transmitters located at (A) but the AF codes in the RDS data enable the receiver immediately to tune to the same programme, but from the main transmitters at (C).

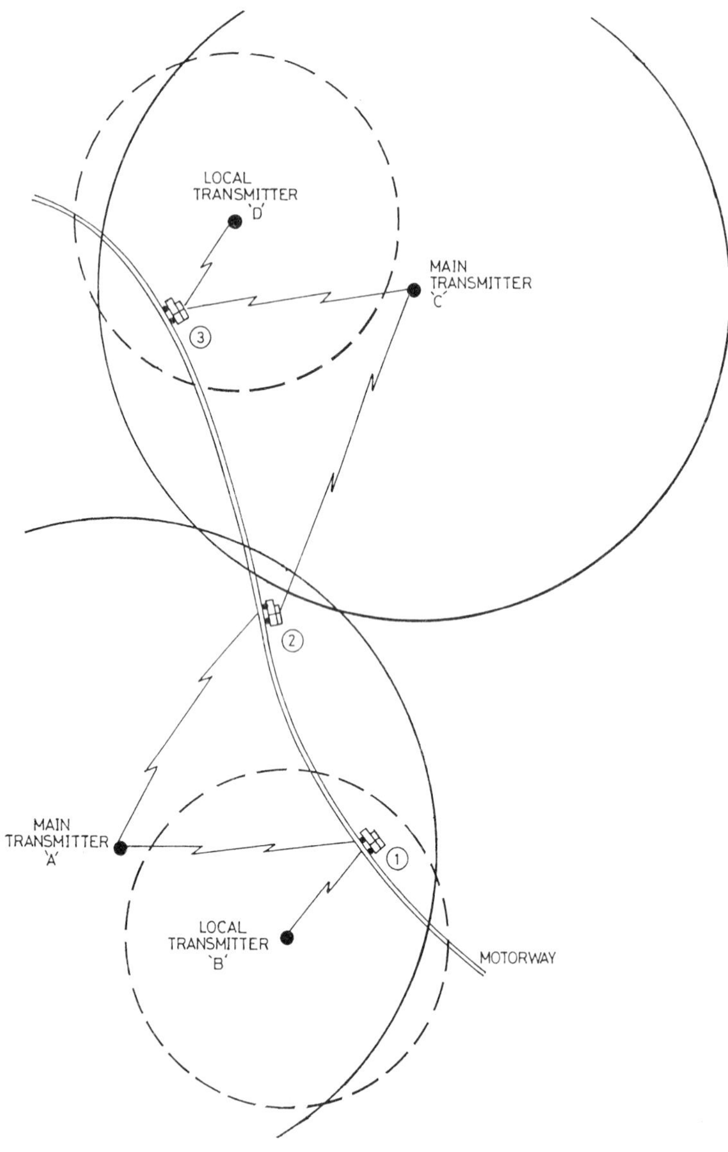

Figure 11.4 Auto-tuning and a traffic information service

The RDS data from all the programmes from the transmitters at (C) again all carry the AF codes for the main transmitters in adjacent areas together with the frequencies of any local transmitters. Again, if a traffic announcement is to be made from the local transmitter (D) covering a different local area the TA flag would be set and the receiver in the car would immediately tune from the main programme of the transmitter at (C) to the local transmitter (D) for the period of the traffic announcement and then switch back to the programme channel as before. The motorist is therefore always able to receive the best transmission for the programme of his or her choice and at the same time is able to receive any traffic announcements covering local areas through which the motorist may be travelling. As the receiver moves from the area of the main transmitter it is automatically reprogrammed with the alternative frequencies for the adjacent areas into which it may move, without any involvement from the user.

The RDS signal is digital and each of these codes represents no more than a single bit in the bit stream. It is thus practicable to repeat the codes frequently in order to give the required rapid response.

Clock time and date

It takes very little of the RDS data stream capacity to transmit a digital code for the time and date. The broadcaster can derive the time information from a national standard, and manufacturers can implement clocks in radio receivers which are truly accurate, requiring no setting and automatically correcting themselves for changes, such as daylight saving time and leap years. They will also incidentally show the time as it is in the country of the broadcast, which could be useful in a journey across national boundaries and time zones.

Clock time and date (CT) consists of a binary coded representation of time and date using Co-ordinated Universal Time (UTC) and calendar date. A local offset is appended in order to allow the RDS receiver to generate an appropriate display of local time as required.

Programme related data

Different codes can be allocated to different types of programme, so that a receiver could search for the user's choice, or display the programme type.

A further code, decoder identification (DI), could be used to switch-in appropriate signal processing, for example noise-reduction. The music/speech code (MS) permits automatic selection between two volume settings, adjusted by the user to suit his or her preferences. The programme identification number (PIN) is an unique code which facilitates unattended recording.

Radiotext and paging

Text messages can be transmitted as radiotext (RT), and the RDS specification gives a format for a 64-character display at the receiver. Applications for radiotext include financial information, sports and news flashes. The specification also allows for the addressing of individual decoders so that such services can be private. The individual decoder address facility may also support a paging service, implemented on an RDS channel. Receivers designed to operate in conjunction with a paging service are similar to conventional fixed frequency pocket pagers, but with automatic station tuning. They may either display simply a telephone number or alternatively a message limited only by the size of the display and the associated processor. The tuning arrangements in such receivers involve a search through the VHF channel to find transmissions which are carrying an RDS signal having a recognizable paging code. The pager identifies the group code address and synchronizes itself in time. The pager then only responds at appropriate predetermined times, typically 6 seconds per minute, in the interest of economizing in battery power. The 'subscriber interface' is to the data management system of the paging service, which feeds the RDS coder located at the transmitter through an RS232 data link. The FM radio signal acts as a carrier and the service can be operated independently of the broadcasting operation.

A transparent data channel (TDC) has been specified and anticipates a requirement for downloading software. The in-house application facility (IH) is intended for the broadcaster's own use, for control or similar purposes.

The RDS specification determines the basic decoder requirements so that receiver decoder manufacturers are able to produce a product, in volume, suitable for a wide range of applications. As has been described, the facilities which are specified are very flexible, to suit the differing requirements of various organizations. A national broadcaster may wish to provide alternative frequencies so that a user in a car, for example, does not have to re-tune as the car moves between different service areas of transmitters. On the other hand, a commercial station may wish to operate a paging or information service, to generate revenue.

RDS networking

Variable data would normally be sent to the RDS coder at the transmitter over telecom links from the appropriate information centre. However, fixed data concerning transmitters is held in the coder in an EPROM.

A re-broadcast (RBS) decoder can be used to take the data derived from the multiplex signal of a receiver and reformat it into an RS232 feed which conforms to the RDS data update specification. This signal is then used as the input to the remote RDS coder. This arrangement assists the broadcaster to 'network' the variable data from transmitter-to-transmitter, avoiding the need for further telecom links.

Data can also be distributed to FM radio transmitters from a television network, using the teletext data channel. As the RDS data rate is quite low it does not impose a very significant overhead on the teletext service. The teletext decoders, located at the FM transmitter sites, decode the data and reformat it as the RS232 drive to the RDS coder. Individual station codes can be used as required so that only specified decoders will respond, thus confining the distribution of the data to specific transmitters.

Appendix: RS 232

The EIA RS232 specification resembles the CCITT V24 specification, the basic difference being in the conventions in naming the connecting pins to be used. Much of the communication between the various units that make up a teletext system are linked with RS232 serial links which use the established signal levels and connections. These are ±5 to 20 volts for data levels, and the connections for a standard 25-way D-type connectors are as follows:

Pin number	*Signal*
1	Protective ground
2	Transmitted data
3	Received data
4	Request to send
5	Clear to send
6	Data set ready
7	Signal ground/common return
8	Received line signal detector
9	+Voltage
10	−Voltage
11	Unassigned
12	Secondary received line signal detector
13	Secondary clear to send
14	Secondary transmitted data
15	DCE transmitter element timing
16	Secondary received data
17	Receiver element signal timing
18	Unassigned
19	Secondary request to send
20	Data terminal ready
21	Signal quality detector
22	Ring indicator
23	Data signal rate selector
24	DTE transmitter signal element timing
25	Unassigned

RS232 connections can cause problems as the specification does not call for all signals to be present. When one piece of equipment requires, say, the 'clear-to-send-signal' (CTS) to be used and the sending end does not support it, the two pieces of equipment will not function when linked together. The solution is to arrange for the missing signals to be substituted by linking the appropriate pin to the required level, either ground or +12 volts, to act as a substitute for the missing signal. This linking is often made between connector pins, and care must be taken that the linking is made at the appropriate end of the cable if all the leads are not present. Such a lead may cause problems when used with a different item of equipment which has a slightly different set of requirements and therefore modified leads should be marked.

The RS232 specification is designed for data over a telephone line, where a modem is usually needed, and the connections are suitably arranged. However, where no modem is used the necessary change of the interconnections is made with a device called a 'null modem' which ensures that the transmit pin of the transmitting device is connected to the receive pin of the receiving device.

Glossary

access time The mean time between selecting a page on a receiver and the first complete reception of that page.

alphanumeric One of the display letters or numbers.

AM Amplitude modulation. A method of modulating an RF carrier with information, its amplitude being changed by the information.

ARI Autofahrer Rundfunk Information A system for transmitting and receiving traffic information used in Germany.

ASCII American Standard Code for Information Interchange.

background colour The colour filling the parts of the character rectangle not occupied by the character itself. The background colour may be any of the seven display colours in level one.

BBC British Broadcasting Corporation.

blast-through alphanumerics Allows upper-class alphanumerics to be displayed while in the graphics mode.

boxed mode The display mode in which the teletext characters appear inset or added to the television picture.

BREMA British Radio Equipment Manufacturers Association.

broadcast service packet The packet which contains information on the broadcast service.

broadcast teletext Teletext intended to be distributed by a broadcast transmission.

bi-phase Digital coding in which the logical ones and zeros are represented by signals of opposing phase.

byte A group of eight consecutive data bits intended to be treated as an entity.

CEEFAX The name of the BBC teletext service.

CCIR International Radio Consultative Committee.

CCITT International Telegraph and Telephone Consultative Committee.

CRC Cyclic redundancy check. A method for checking for errors by counting the digits in a section of code and comparing that number with the original number inserted into the code when the code was originated.

An additional code added to the transmission (the CRC code) which relates in a mathematical way to a known sector of the code in the message. Repeating the mathematical process at the receiver and comparing the result with the CRC code allows the validity of the message to be confirmed.

character byte The byte obtained by appending the parity bit to the *character code*.

character code A 7-bit binary number representing one of the display characters or control characters.

character rectangle One of the 960 units in the regular matrix of 24 rows by 40 sites in which the characters are displayed on the screen.

character row See *Row*.

CTS Clear-to-send.

clock run-in A sequence of bits at the start of a data line to allow a receiver to achieve bit synchronization.

co-channel Interference from a distant station on the same nominal carrier frequency.

command row The row usually at the bottom of the screen, which is used to send commands to the teletext system from a terminal. These rows are not displayed as part of the page.

conceal A display mode during which designated characters, although stored in the decoder, are displayed as spaces until the viewer chooses to *reveal* them by a command from the handset.

contiguous graphics set The set of 96 display characters comprising the 64 contiguous characters together with the 32 blast through alphanumerics characters.

contiguous mode The display mode in which the six cells of the graphics characters fill the character rectangle.

control bits Each page header contains 11 control bits to control the display of the page.

control character One of the 32 characters that are not displayed but control the character or page display modes.

current loop Method of interconnecting terminals and transmission equipment where a '1' is indicated by a current and a '0' is indicated by the absence of current.

data amplitude The separation between the all 0s level and the all 1s level of the data signal. In the World System Teletext intended for use on 625-line television systems, the data amplitude is specified to be 66% of the nominal video level.

data line One of the otherwise unused lines in the field blanking interval (FBI) of a television signal that may be used to carry teletext information.

DBS Direct broadcast by satellite.

DCM Decoding margin meter, an instrument for measuring the eyeheight of a data signal, taking into account the noise level present.

display character One of the shapes that can be generated for display in character rectangle as part of the page.

display colour The colour used to depict a character against the background colour in a character rectangle. At level 1 of WST, this may be one of seven colours, but with higher levels many shades of colour can be reproduced.

display mode The way in which the character codes corresponding to display characters are interpreted and displayed.

DRCS Dynamically redefinable character sets.

DSR (data set ready) A modem interface control signal indicating to the attached terminal that the modem is connected to the telephone circuit.

DTE (data terminal equipment) The equipment acting as the data source.

DRT (data terminal ready) The modem interface signal indicating to the modem that the terminal is ready to transmit.

EBU European Broadcasting Union.

eyeheight In a noise-free signal, the smallest difference between any 'l' data pulse and any '0' pulse for sampling positions equally spaced at the data rate and position chosen to maximize the quantity. It is expressed as a percentage of the data amplitude.

FBI Field blanking interval. In a television signal, the interval between the fields of video information. It is now preferred to 'VBI', vertical blanking interval.

flash A display mode in which the display of a character(s) alternates with the display of a space(s) under the control of a timing device in the decoder.

FLOF Full level one features, a term used during the development of level 2 WST.

FM Frequency modulation, a modulation method where the frequency of the carrier signal is changed in response to the modulating signal, the carrier amplitude being kept constant.

foreground colour The colour of alphanumerics or graphics.

framing code A byte following the clock run-in sequence which allows the receiver to achieve byte synchronization. In WST it is so protected that it may be currently decoded even if one of its bits is wrongly decoded.

graphics character One of the 64 different display characters based on the division of the character rectangle into six cells. The cells can be displayed contiguously or separated.

graphics mode The display mode in which the display characters are those of one or the other graphics sets depending on whether the contiguous or separated mode is being used.

Golay code A forward error connecting code. In the (23 12) form used in the MAC/packet system, Golay code allows for the correction of up to three errors in the group of 23 bits.

ghost-rows An early term for an additional row that was not intended to be displayed but which contained additional control information for the page.

Hamming code A forward error-correcting code, due to Hamming. Several versions are used in WST. At level 1 the Hamming code is a byte containing four message bits and four protection bits. Hamming codes are used to protect address or control information and, in the stated form, are able to correct one error in the byte.

handshake An exchange of signals used in communication protocols to indicate that the previous part of the signal had been received successfully and that the next part can be sent.

hold graphics A mode of graphics display in which any control character occurring during the graphics mode results in the displayed graphics character being held to cover the space occupied by the next control character. Its use allows the avoidance of space which would otherwise occur between changes of graphics colour.

IBA Independent Broadcasting Authority.

IRT Institut für Rundfunktechnik (Germany).

linked pages Pages which contain information which 'link' them to another page. Packet 27 is designed to carry information of this type.

LSI Large scale integrated circuit, a complex device in which a large number of semiconductors are deposited on one piece of silicon usually to perform a particular task.

magazine A group of up to 100 numbered pages, each carrying a common magazine number in the range 1–8. Up to eight magazines may be transmitted in series or parallel on a television channel.

network identification Part of the broadcast service packet 8/30.

newsflash page A page in which all the information for display is boxed so that it can be seen better against the television picture. Control bit C5 is set to indicate to the decoder that the information should be inset into the television picture or added to the television picture.

NRZ Non-return-to-zero, a form of coding for data in which the data stays in the 'one' position until the next zero data pulse causes it to return to zero.

ORACLE The name of the UK independent television companies teletext service.

page A group of up to 24 rows of 40 characters intended to be displayed as an entity on the television screen.

page header A page header data line has a row address 0 and it separates the pages of a magazine in the sequence of transmitted data lines. In place of the first eight-character byte, it contains Hamming coded address and control information relating to the page. Thus the top row of the page has only 32 display character bytes. These are used for the transmission of general information such as magazine and page number, day and date, programme source and clock time.

page subcodes An addition to the page numbers to extend the page address range.

PDC Programme Delivery Control, the control of domestic videocassette recorders by teletext packet 8/30.

RDS Radio Data System, a method of adding a sub-carrier signal to the normal FM stereo radio signal for the transmission of digital data.

release graphics The display mode in which control characters are invariably displayed as spaces. It is complementary to the hold graphics mode.

reveal The display mode complementary to the conceal mode.

rolling headers The use of the top row of the page to display all the page headers of the selected magazine as they are transmitted. This gives an indication to the viewer of the page transmission sequence, watching or awaiting a selected page.

RS232-C Electronic Industries Association (EIA) standard applicable to the interconnections between data terminal equipment (DTE) and data communication equipment (DCE) employing serial communications. The 'C' indicates the latest revision and the standard is usually referred to as RS232. (See Appendix for application details.)

row A page comprises 24 rows of characters. When displayed on a television screen each row occupies about 20 television lines. To avoid confusion with 'television lines' the lines of text are called rows.

row adaptive transmission Teletext transmission in which rows containing no information are not transmitted rather than being transmitted as a row of space characters. This reduces the access time of the transmission. The non-transmitted rows are displayed as unboxed black spaces.

separated graphics set The set of 96 display characters comprising the 64 separated graphics characters (corresponding to the contiguous graphics characters) together with the 32 alphanumeric characters.

separated graphics mode The display mode in which there is background colour boundary around and between the six cells of the graphic characters within the character rectangle.

smoothed graphics A graphics set available in the higher levels of WST, giving better resolution graphics.

space A character rectangle entirely filled by the background colour.

sub-carrier An additional carrier signal for transmission of supplementary information.

subtitle page A page in which all the information for the display is boxed and the control bit C6 is set to allow the decoder to automatically inset or add to the television picture.

teletext A method of transmitting data, usually in the field blanking interval of a television signal.

television data line *See* data line.

time coded page A page containing a time code in the header, and transmitted at a pre-set time.

time display The last eight characters of every page header are reserved for clock-time. A receiver may be arranged to display these characters to give a clock-time display as an insert into the picture.

VBI *See* FBI.

videotext This is the general name used for an information service that uses a telephone line for the transmission of the data.

WST World System Teletext.

References

1. *Specification of standards for information transmission by digitally coded signals in the field-blanking interval of 625-line television systems.* Joint BBC, IBA and BREMA publication, October 1974
2. Hofmann, H. and Lau, A. Results of the propagation tests with teletext signals according to the British Standard, over the ARD and ZDF television transmitters, *Rundfunktechnische Mitteilungen* (23) 25–36, 1979
3. *Broadcast Teletext Specification.* Joint BBC, IBA and BREMA publication, September 1976
4. Teletext and viewdata, *Proc. IEE,* **126**, 1349–1428, December 1979
   (a) Mothersole, P. L. Teletext and viewdata: new information systems using the domestic television receiver
   (b) Clarke, K. E. International standards for videotext codes
   (c) Betts, W. R. Viewdata: the evolution of home and business terminals
   (d) Pandy, K. Second generation teletext and viewdata decoders
   (e) Beakhurst, D. J. and Gander, M. C. Teletext and viewdata: a comprehensive component solution
   (f) Hutt, P. R. and McKenzie, G. A. Theoretical and practical ruggedness of UK teletext transmissions
   (g) Rogers, B. J. Methods of measurement on teletext receivers and decoders
   (h) Green, N. Sub-titling using teletext service: technical and editorial aspects
   (i) Hedger, J. and Easton, R. Telesoftware: adding intelligence to teletext
   (j) Crowther, G. O. and Hobbs, D. S. Teletext and viewdata systems and their possible extension to the USA
   (k) Chambers, J. P. Teletext: enhancing the basic system
5. Hamming, R. W. Error detecting and error correcting codes, *Bell System Technical Journal,* **29**, 147–160, 1950
6. *DIDON-ANTIOPE specifications techniques,* Télé Diffusion de France, 1984
7. Gecsei, J. *The Architecture of Videotex Systems,* Prentice-Hall Inc, Englewood Cliff, NJ, 1983
8. Storey, J. R., Vincent, A and Fitzgerald, R. A description of the broadcast teleidon system, *IEEE Transactions on Consumer Electronics,* **CE-26** (3), 1980
9. Developments in teletext, *IBA Technical Review,* May 1983
   McKenzie, G. A. Teletext – the first ten years
   Crowther, G. O. Teletext enhancements – levels 1, 2, 3 and 4
   Vivian, R. H. Level 4 – teletext graphics using alpha-geometric coding
   Johnson, G. A. and Slater, J. N. The networking of ORACLE
   Mothersole, P. L. Equipment for network distribution
   Staff at the Mullard Applications Laboratory, Integrated circuits for receivers
   Lambourne, A. D. Newfor – an advanced subtitle preparation system
10. Kinghorn, J. R. New features in world system teletext, *IEEE Transactions on Consumer Electronics,* **CE-30** (3), 1984

11. Tarrant, D. R. Teletext for the world, *IEEE Transactions on Consumer Electronics,* **CE-32** (3), 1986
12. Crowther, G. O. Adaptation of UK teletext system for 525/60 Operation, *IEEE Transactions on Consumer Electronics,* **CE-26** (3), 1980
13. *World System Teletext and Data Broadcasting System, Technical Specification,* due to be published by the CCIR in 1990; obtainable from BREMA (B. J. Rogers), Landseer House, 19 Charing Cross Road, London WC2H 0ES
14. *North American Broadcast Teletext Specification (NABTS),* CBS Television Network, June 1981
15. *ARD/ZDF/ZVEI Directive 'TOP' System for Teletext,* Institut für Rundfunktechnik, Munich
16. *Specification of the Domestic Video Programme Delivery Control System (PDC),* European Broadcasting Union, Technical Centre, Geneva
17. The D-MAC packet system for satellite and cable, *IBA Technical Review,* November 1988
    Windram, M. D., Tonge, C. J. and Hills, R. C. *Satellite Broadcasting in the UK.* (An overview of the D-MAC packet system. The vision signal. Sound channels. Interpretation blocks. Service identification signals. Data broadcasting. High priority information – line 625. The conditional access system: RF transmission of D-MAC/packet.)
18. *Specification of the MAC/packet family, EBU Document, Tech. 3258-E,* October 1986
19. Lender, A. The duobinary technique for high-speed data transmission, *IEEE Transactions on Comms. Electronics,* 66, 214–218, 1963
20. Macdiarmid, I. F. A testing pulse for television links, *Proc. IEE,* **99**, Part IIIA (18), 436–444
21. Spicer, C. R. and Tidy, R. J. *An Instrument for the Automatic Measurement of Teletext Transmission Quality,* IERE Television Measurements Conference Publication, 1979
22. Sherry, L. A. and Hills, R. C. The measurement of teletext performance over United Kingdom television network, *The Radio and Electronic Engineer,* **50**, 503–518
23. Mason, A. *DELPHI, A Versatile and Controlled Teletext Signal Source,* IBC Convention Publication, 253–257, 1978
24. Chambers, J. P. BBC datacast, *EBU Review–Technical No. 222,* April 1987
25. *Specification of the Radio Data System, RDS, for VHF/FM Sound Broadcasting,* Specification Tech. 3244-E, European Broadcasting Union, Technical Centre, Geneva, 1984
26. *Guidelines for the Implementation of the RDS System,* Document Tech. 3260, European Broadcasting Union, Technical Centre, Geneva, 1990

Index